Exposing Deep State Science Mistakes, Lies and Environment Devastation

J. Marvin Herndon, Ph.D.

Copyright © 2024 J. Marvin Herndon, Ph.D.

All rights reserved.

ISBN: 9798304787789

FOREWORD

The 2024 US presidential election was unlike any other in recent history. American voters repudiated the Deep State and its concomitant criminal activity [1].

The term Deep State refers to the cabal of globalists, elites, multinational organizations and corporations, government officials and employees, and others who collectively operate as a "shadow government" to usurp and subdue the will, freedoms, health, and prosperity of ordinary people and sovereign nations.

Americans and others throughout the world are beginning to realize that they have been lied to and deceived by their governments, mainstream media, and others. Famously, US CIA Director (1981-1987) William Casey reportedly said, "*We'll know our disinformation program is complete when everything the American public believes is false*" [2]. Fortunately, people are starting to wake up to the orchestrated deceit around them, realizing that they are constantly being lied to by main stream media and that their Internet information is being sieved into false representations of the real world. However, few understand the depth of the deceit; most they have a long way to go. And, it is not harmless lies, but lies about covert activities that pose serious threats to life on this planet.

Science is all about truth. Over my professional lifetime of half a century, I have witnessed science in America, the European Union, and the British Commonwealth increasingly abrogating long-held scientific standards, and becoming flawed and fraught with Deep State lies. I can only attest to those areas in which I have had experience, namely, Earth and planetary sciences, astrophysics, and environmental sciences.

The Deep State has co-opted legitimate science by failing to cite (or refute) new advances, to lie, and to engage in non-science methodologies. Recall these words: *And ye shall know the truth, and the truth shall make you free* (John 8:32).

Exposing Deep State Science

CONTENTS

	FOREWORD	iii
	ACKNOWLEDGEMENTS	vii
1	WHAT CONSTITUTES GOOD SCIENCE	1
2	EARTH SCIENCE MISTAKES AND DECEIT	11
3	PLANETARY SCIENCE MISTAKES AND DECEIT	57
4	ASTROPHYSICAL SCIENCE MISTAKES AND DECEIT	97
5	DEEP STATE ENVIRONMENT DESTRUCTION AND DECEIT	115
	REFERENCES AND AUTHORITIES	167
	INDEX	225
	ABOUT THE AUTHOR	233

ACKNOWLEDGMENTS

The following individuals contributed in major ways to my understanding of science: Paul K. Kuroda, Inge Lehmann, Lynn Margulis, Marvin W. Rowe, Hans E. Suess, and Harold C. Urey. I gratefully acknowledge the following individuals who contributed as coauthors cited in Chapter 5: Ian Baldwin, Mark Hagen, Raymond D. Hoisington, Mark Whiteside, and Dale D. Williams.

1 WHAT CONSTITUTES GOOD SCIENCE

Observations, ideas, and understandings are the substance of science. We are in a very real sense creatures of the mind building science upon nothing more tenuous than the fabric of human thought, a fabric that must be passed down from generation to generation without unraveling.

In 1974, only a few months after receiving the Ph.D. in nuclear chemistry, I presented a seminar at the University of California, San Diego. There were two men in the audience whom I only knew by reputations: One, Noble Laureate Harold C. Urey (1893–1981), had discovered deuterium, heavy hydrogen of mass two, and had conceived the idea of using oxygen isotopes to determine temperature in times past, called paleothermometry. The other, Prof. Dr. Hans E. Suess (1909–1993), co-discovered the shell structure of the atomic nucleus which earned J. Hans D. Jensen a share of the Nobel Prize (Figure 1.1).

Figure 1.1. Harold Clayton Urey (left) and Hans Eduard Suess (right).

Both Urey and Suess were recipients of knowledge passed down from masters. Urey had served a post-doctoral apprenticeship with Niels Bohr in Copenhagen. Suess had learned geology from his father Franz Eduard Suess, a famous geologist, who had learned from his father, Eduard Suess, an even more famous geologist and author of *Das Antlitz der Erde* [3]. Something that I said during that seminar led to my being invited to serve a post-doctoral apprenticeship to these two senior giants.

Suess and Urey were well schooled in the principles, methods, and ethics of pre-WWII science, a time of little government funding. In 1950, the US National Science Foundation was established and wrote the rules for government administration of science-research funding that today permeate the scientific community. Sadly, however, these rules were conceived without considering human nature. For example, the rules included secret funding-proposal reviews by one's competitors which encourage deceit, as well as proposal requirements that trivialize science. How can one specify beforehand what will be discovered that has never before been discovered and what one will have do to make that discovery?

By 1974, the fabric of science was already being frayed. Now, 50 years later, I wish to share some of the insights I learned from Urey and Suess and also picked up along the paths of making scientific discoveries, for example [4-33].

The purpose of science is to determine the true nature of Earth and Universe and all contained therein. The word "true" is paramount. Science is all about truth and integrity. But in many other activities, politics for example, truth does not have the same necessity as it does in science, although as acknowledged by Mahatma Gandhi, *"Truth never damages a cause that is just."*

Science is the ever-evolving activity of replacing less-precise understanding with more-precise understanding. In this way, science advances. But how does one know, for example, whether a new idea represents an advance or not? How does one determine the truth in such an instance? In mathematics one can prove that which is true, but generally not so in science.

When a new idea comes along there should be discussion and debate. If possible, efforts should be made to refute the new idea, to show that it is *not* true. If the scientific community is unable to refute the idea, ideally in the same journal where it was first published, then the idea should be acknowledged and cited in relevant scientific literature. Beware of the science-charlatans that ignore contradictory new ideas, misleading the public and other scientists, and cheating those who fund their work, mainly taxpayers.

The criterion for truth in science is different than for truth in other fields. Jurisprudence, for example, filters evidence as to whether admissible or inadmissible and allows a jury to determine truth, i.e., guilt or innocence, which may or may not be the actual fact. In matters of political governance, for example, consensus is the criterion for truth, but in science, consensus is nonsense; science is a logical process, not a democratic process.

Fundamental new ideas sometimes meet with resistance. There is, I have observed, a human analogue to Lenz's Law in physics and Le Chatelier's Principle in chemistry, the tendency of a system to oppose change. On one occasion after a pleasant dinner, I began to explain my recent discoveries to a friend, a visiting scientist whom I had not seen for several years. As I described how Earth's interior differed from what he had been taught, his demeanor changed, his face became ashen; he hardly spoke the remainder of his visit. I have encountered similar experiences with others.

In 1623, Galileo Galilei (1564–1642), one of the greatest scientists of the millennium, precisely characterized human response to new ideas in a letter written to Don Virginio Cesarini, stating in part, *"I have never understood, Your Excellency, why it is that every one of the studies I have published in order to please or to serve other people has aroused in some men a certain perverse urge to detract, steal, or deprecate that modicum of merit which I thought I had earned, if not for my work, at least for its intention"* [34].

When I am exposed to a fundamentally new concept, I ask myself, *"Suppose the new concept is correct? What does it mean? What advances might follow from it?"* Therein might be opportunities for new discoveries.

Good science, properly executed and securely anchored to the abundances of the elements and to the properties of matter and radiation, transcends human opinion. Ideally, one seeks to derive fundamental quantitative relationships in nature.

When I began my postdoctoral apprenticeship, university professors and their students were first beginning to make computational models. Models are computer programs that generally begin with an assumed end result which is then attained by result-selecting variables and assumptions. The making of models that are based upon assumptions, on the other hand, in my view is generally not science. A few models are useful [35], for example those aimed at predicting the paths of hurricanes, but models do not generally lead to scientific discoveries.

I had just begun a three-year postdoctoral apprenticeship with Hans E. Suess and Harold C. Urey when, in the morning of the third day, Suess stopped by my office , handed me a reprint of one of his scientific papers to read, and asked if I would like to stop by his office later and discuss it with him. Wanting to make a good impression, I read the paper quite carefully. It seemed simple enough, almost trivial. For good measure, I re-read it and then went to his office.

Not five minutes of discussion had taken place before it became painfully evident to me that I had completely failed to understand the paper, which contained neither complex mathematics nor necessitated specialized background information. Suess just shook his head and told me to come back when I understood it.

I was devastated. I had really wanted to make a good impression. Dejectedly, I left Suess' office to meet Harold Urey for lunch. Urey sensed that something was wrong and asked for an explanation. I explained the impossibility of understanding Suess' paper. Urey then smiled kindly and suggested that I might try reading scientific articles the way he does. Urey explained that he reads only one sentence and does not progress to the next until he understands fully the meaning of that one sentence. I put into practice Urey's suggestion, and it was as if a whole new world had opened up to me – I could understand Hans Suess' scientific papers just as he had intended them to be understood.

So, in the first week of my postdoctoral apprenticeship, I had learned how to read, but had not yet realized that I also needed to learn how to write in logical, causally related steps. That would come two months later.

One afternoon six months into my post-doctoral apprenticeship, Suess asked me directly if I knew why he had chosen me. Then he reminded me of my seminar and the questions that followed and one specific question in particular, which I had long since forgotten. He reminded me that I had answered by saying that I could not answer that question, that the information was simply not known. Looking at me with a gaze that seemed to stare into my soul, Hans Suess told me that not one young scientist in a thousand would have answered that way; most would have tried to answer the question. Then, he explained that it is much more important to know what is *not* known, than to know what *is* known.

There is a technique, a methodology, one can apply to begin to know what is not known and that is quite simply to go back in time [36]. Travel through time, not with a H. G. Wellsian time machine, but through a historical understanding of the events and ideas that led to the present state of understanding. All of that is documented in the scientific literature. Logically ordering historical observations and ideas into a sequential progression of understanding, while being keenly aware of later changes and discoveries, helps one to see gaps in the sequence, to begin to know what is not known, and perhaps to find mistakes that were made and not corrected in light of subsequent data.

Division and progressive subdivision with specialization comprise an integral process in nature and in human activity. Indeed, each of us began as a single cell which divided and progressively subdivided while achieving specialized functions. Ever-burgeoning observation, experimentation, derivation, calculation, and understanding, out of necessity, have led to division, progressive subdivision, and specialization of knowledge. By the 17th Century, chemistry was developing its distinction as a clearly separate science from physics. Then, in the 20th Century, as academicians expanded study of the Earth, those same divisions were carried forward as geochemistry and geophysics.

But there is a problem: As geochemistry and geophysics are only partial descriptions of the Earth, their separation and specialization poses a serious impediment to understanding, and, consequently, to making important new discoveries, particularly in instances when geochemists have little training in physics and when geophysicists have little training in chemistry. Another, sometimes even more serious impediment to making important new discoveries, and one often least appreciated, arises as a consequence of excluding, from the realm of scientific investigation, understanding of relevant science history.

Science is very much a logical progression through time. Advances are frequently underpinned by ideas and understandings developed in the past, sometimes under circumstances which may no longer hold the same degree of validity. It is of great benefit for a scientist, working within a conceptual framework, to understand the historical basis of that framework, to understand how the present state of knowledge arose and under what circumstances.

All too often, scientists, being distinctly human creatures of habit, plod optimistically along through time, eagerly looking toward the future, but rarely looking with question at circumstances from the past which have set them upon their present courses. Progressing along a logical path of discovery is rather like following a path through the wilderness. Occasionally, one comes to a juncture, the path splits, presenting a choice of scientific interpretations. Choose the correct logical interpretation and the way is clear for further progress; the wrong choice leads to confusion. That is often the way of science. To make matters even more complicated, the correct path is sometimes invisible, obscured because some requisite discovery has not yet been made. Moreover, the logical progression of scientific discovery is often opposed by the darker elements of human nature and institutional self-interest.

Much has been written about the Roman Catholic Church's opposition to the heliocentric hypothesis of Nicolaus Copernicus (1473–1543) [37] and its consequences on individuals and on the progression of human knowledge [38]. Less known, though, is that about 1800 years before Copernicus, Aristarchus of Samos (310–230 BC) had arrived at the same idea. Although the original explanatory document is lost, clear reference is given to his ideas by Archimedes (287–213 BC) in his book *The Sand Reckoner* which states in part, *"His hypotheses are that the fixed stars and the sun remain unmoved, that the earth revolves about the sun in the circumference of a circle, the sun lying in the middle of the orbit, and that the sphere of fixed stars, situated about the same center as the sun ..."* [39].

What, one might logically ask, is the relevancy of the above historical references, especially now in the time of near-instantaneous global communications and Internet access? The relevancy relates to the persistence of human nature, which does not change on a time-scale of a few hundred or even a few thousand years, and which underlies impediments posed by institutional self-interest.

Phenomena, processes, or events, when described in terms of a problematic paradigm, yield explanations that are generally more complex, if not logically unrelated or physically impossible, than corresponding explanations posed later within a different, better understood, and more-correct paradigm. For example, in the Ptolemaic Earth-centered universe paradigm, the observed apparent motion of planets, specifically their retrograde motions, were described by complex epicycles (Figure 1.2). Within the state of knowledge at the time, that explanation seemed to explain the observed retrograde planetary motions, but we now know that epicycles are artificial constructs and that Earth is not located at the center of the Universe.

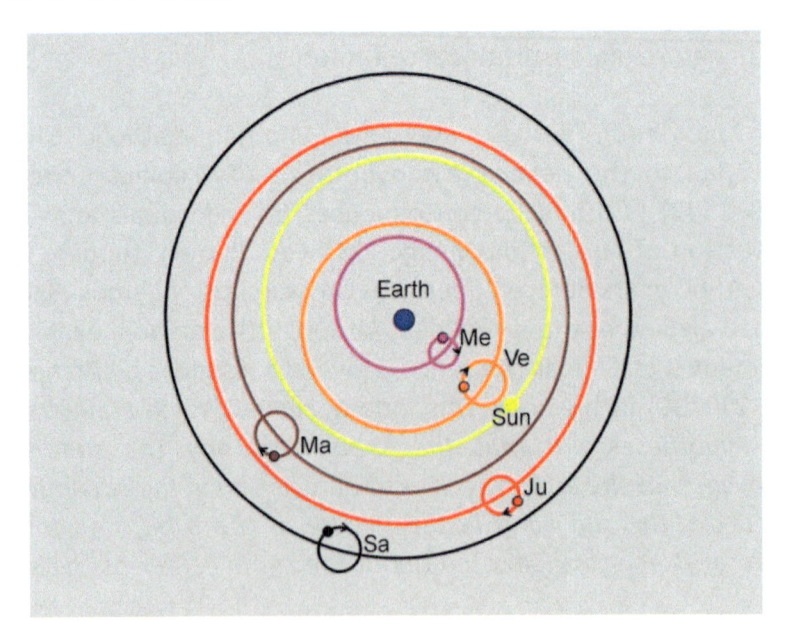

Figure 1.2. Epicycles were able to explain apparent retrograde motion of planets in the problematic Earth-centered Ptolemaic universe paradigm.

For another example, in plate tectonics theory mountains are thought to form by plate collisions [40], as plates move about the globe riding atop assumed mantle convection cells. Within that belief, mountains older than Pangaea required an earlier continent formation and breakup, and then an earlier one, etc. In other words supercontinent cycles, also called Wilson cycles [41] (Figure 1.3).

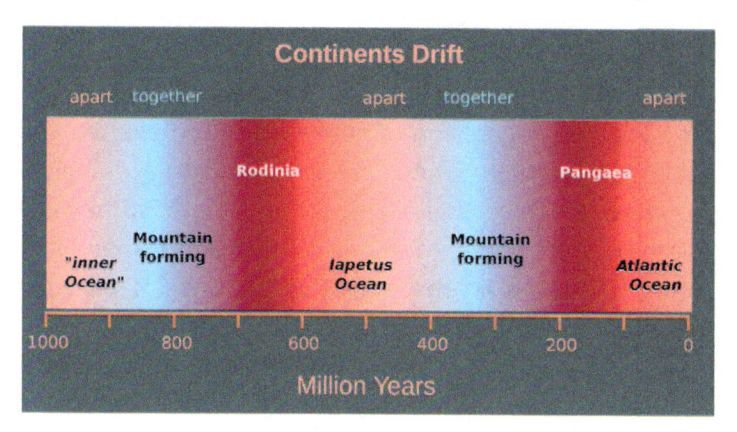

Figure 1.3. Illustration showing the fictional plate tectonics idea of supercontinent cycles. Courtesy of Hannes Grobe.

The lesson to be learned is this: If complex *ad hoc* explanations are necessary to make some observations seem to fit within current knowledge, then consider that as an invitation to question current knowledge.

Similarly, in the classical, pre-quantum physics paradigm, an ideal black body in a state of thermal equilibrium was calculated to emit radiation with essentially infinite power in the shorter wavelengths. This is the so-called ultraviolet catastrophe, a circumstance that is physically impossible. Later, in the now-known, more-correct quantum physics paradigm, black body radiation, and other phenomena, can be explained logically, causally, and with greater simplicity, without invoking complex, *ad hoc* assumptions. Such a fundamental change in understanding is referred to as a *paradigm shift* [42].

Science is like a long road paved with observations, ideas, and understandings. From a distance it might seem like a smooth strip of ribbon meandering through time. But up close, it can be seen as a rocky road indeed — a mix of insight and oversight, design and serendipity, precision and error, and implication and revision, all too often influenced by the vagaries of human behavior. By considering deeply the relevant science history, one might begin to recognize past faltering in the logical progression of observations and ideas and, perhaps then, to discover new, more precise understandings [36].

Science is a logical progression of causally related events, analogous to a really good movie where all the actions are logically and causally related; the pieces all fit together. Now, if something about nature seems like a really bad movie and does not make sense, ask the question, "What is wrong with this picture?" That can be the first step toward making an important discovery.

There is a more fundamental way to make discoveries than the variants of the scientific method taught in schools which I describe here: *An individual ponders and through tedious efforts arranges seemingly unrelated observations into a logical sequence in the mind so that causal relationships become evident and new understanding emerges, showing the path for new observations, for new experiments, for new theoretical considerations, and for new discoveries* [36].

Science should not simply be an academic discipline, but should aim to improve the well-being of life on Earth. By virtue of their abilities and training, scientists in my view have a special responsibility to humanity, not only to improve human and environmental health, but to protect life on this planet. Life on Earth is possible due to both the nature of Earth's composition and physical processes, which afford protection from the ravages and variations of solar radiation, and the myriad complex interactions by and between biota and their various environments. Scientists should avoid and indeed prevent any activity that upsets the delicate balance in nature.

Above all, scientists must be truthful..

2 GEOSCIENCE MISTAKES AND DECEIT

Albert Einstein [43] worked diligently, but unsuccessfully, to understand the origin of Earth's magnetic field, which he considered to be one of the five most important unsolved problems in physics [44].

Although the magnetic compass was used in antiquity [45], the cause of its operation, now referred to as the Earth's magnetic field or geomagnetic field, was a great mystery. For centuries, it was not known whether that magnetic field originated within the Earth or was extra-terrestrial in origin. In 1600, William Gilbert [46] showed that magnetic compass deflections measured around a sphere fabricated from magnetic loadstone corresponded to compass deflections recorded by navigators around the surface of our planet. In 1838, Carl Friedrich Gauss [47] showed mathematically that the seat of the geomagnetic field resides at or near Earth's center.

In 1855, Michael Faraday [48] reported his discovery that an electric current, i.e., the flow of electric charges, produces a magnetic field. Earth's fluid core, discovered in 1906 [49], was the only interior region thought to be capable of motion in addition to planetary rotation, until 1993 [19]. In 1939, Walter Elsasser [50] suggested that the geomagnetic field might be produced by convection in Earth's fluid core, which, coupled with planetary rotation, acts as a dynamo, a magnetic amplifier. Elsasser [50-52] simply assumed that convection exists in the fluid core without any independent corroborating evidence. More than 80 years later no independent corroborating evidence has been discovered.

Life on Earth depends critically on the geomagnetic field, which not only serves as a navigational aid for many creatures [53, 54], but also acts as a shield that protects all life on Earth from the charged-particle rampages of the solar wind [55] (Figure 2.1). From time to time, massive pulses of charged plasma are ejected from the sun's corona and partially overwhelm Earth's magnetic field [56]. During these times, charge particles stream through Earth's atmosphere lighting the Northern and Southern skies with dazzling auroral displays. These sporadic events induce dangerous electrical currents in long metallic conductors at the surface and damage electrical equipment by induced electric currents [57, 58]. Although infrequent, these events prefigure the potential calamities that will inevitably occur when the geomagnetic field weakens, reverses and/or collapses [16].

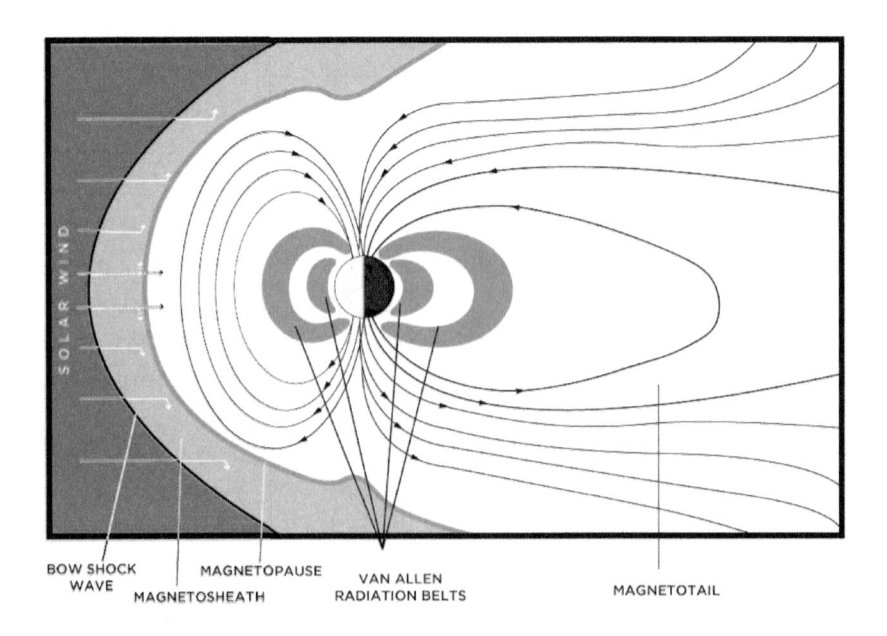

Figure 2.1. Schematic representation of the geomagnetic field, Earth's shield that deflects charge particles from the sun safely away from the planet. From [59].

From time to time irregularly, the geomagnetic field reverses. The last geomagnetic reversal occurred about 786 thousand years ago. Our stone-age ancestors with no technological infrastructure survived that reversal. During the next geomagnetic field reversal and/or collapse the charged-particle onslaught from the sun will ravage our electrically-based infrastructure, potentially wiping out two centuries of infrastructure development [60]. One of the foremost obligations and responsibilities of scientists should be to advance geomagnetic understanding to protect humanity. Instead, as described below, the global geoscience community, functioning as a cartel, for decades has systematically deceived world governments, scientists, and the public about the origin and nature of the geomagnetic field and its potentially near-term risks to humanity's infrastructure [16].

Consensus Nonsense: In 2020, Li et al. published an article entitled *"Shock melting curve of iron: A consensus on the temperature at the Earth's inner core boundary"* [61]. Its title illustrates well the non-science that permeates the geoscience establishment, specifically, the failure to understand principles of science, which are contrary to consensus, and the nature of Earth's interior.

In the realm of politics, consensus is a measure of the popularity of an idea, not necessarily its correctness. In science, consensus is nonsense. At the frontiers of understanding, at the interface of the unknown, the popularity of a concept in science is not a measure of correctness. Providing data that supports a consensus is not the way science progresses; if consensus determines scientific truth, progress would be impossible. Scientific paradigm-shifting revolutions inevitably overturn consensus. Science progresses by determining what is wrong with currently-held perceptions.

When a new concept emerges in science that challenges important current thinking, the obligation of the scientific community is to attempt to refute the new concept. If unable to do so, the new concept should be cited in subsequent literature. To ignore or fail to cite a new concept that challenges important current thinking, is not only poor science, but it cheats those who fund the research, usually taxpayers, and cheats fellow scientists who might otherwise make advances on the new concept.

Li et al. [61] join hundreds, if not thousands, of scientists who for forty years have systematically ignored or failed to cite published contradictions to the very consensus they attempt to support. That collective failing has wasted millions of dollars of taxpayer-provided research funding and has misled government officials, scientists, and the public into a false sense of security concerning the risks and consequences of a geomagnetic reversal and/or collapse. Why? To what end? Malfeasance by Deep State government funding-agencies and scientific publishers are partially to blame, but so too are the scientists, some of whom are corrupt, or are afraid to speak out, or are simply ignorant of what constitutes good science.

New Deep-Earth Paradigm Shifts: In the following I present a historical record of the deceitful response to a challenging new concept published in 1979 [4]. Not only has that challenging new concept been systematically ignored, but concerted efforts have been made to deceive the public of its consequential advances, many of which are related to geomagnetic field origin [6, 13, 19, 20, 62] and the next potential geomagnetic reversal and/or collapse [16].

In 1906, Oldham discovered Earth's iron metal core whose boundary lies about half way to the planet's center [49] (Figure 2.2). By 1930, its dimensions were well established and the core was found to be liquid [63]. A simple picture of Earth's interior emerged: An iron alloy core surrounded by a silicate-rock mantle and topped with a very thin crust (discovered by Mohorovičić in 1909 [64]). But something was missing. Earthquake waves from a large New Zealand earthquake, instead of being shadowed by the core, were actually observed at the surface in the shadow zone. This posed a great geoscience mystery.

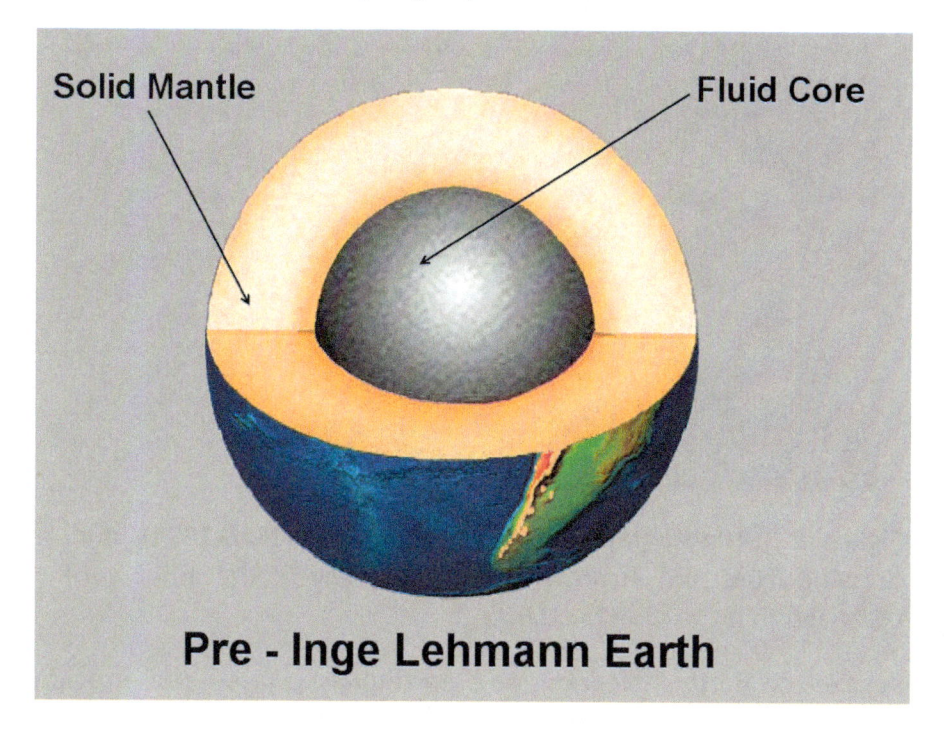

Figure 2.2. The simple picture of Earth's interior as understood in 1930.

In 1936, the Danish seismologist, Inge Lehmann, solved this great mystery by correctly deducing that within the fluid core there must be a solid inner core that would reflect earthquake waves into the shadow zone, thus explaining seismic observations [65]. Figure 2.3 shows her discovery diagram.

Lehmann's reasoning was of such great precision that her inner core concept was accepted as fact even though confirmatory evidence was not available until the 1960s.

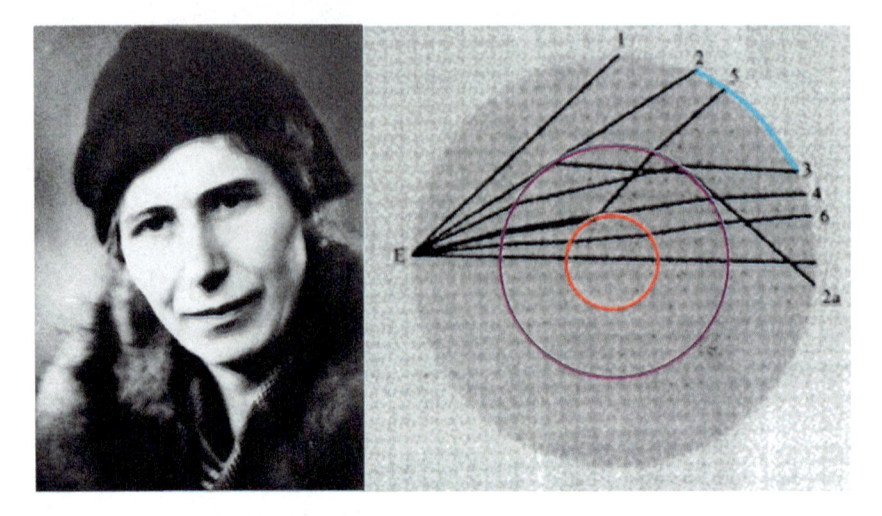

Figure 2.3. Photograph of Inge Lehmann (1888-1993) and a drawing from [65] illustrating her discovery of the inner core. I colorized that drawing for clarity.

Studies of Earth's rotation and earthquake waves can provide information on the distribution of mass-layers within the planet. The chemical composition of those layers, however, must be deduced from studies of meteorites. In the 1930s and 1940s, Earth was thought to resemble an *ordinary chondrite* meteorite, called ordinary because of their great abundance. If heated sufficiently in the laboratory, the elements of an ordinary chondrite separate into two components, an iron alloy beneath silicate-rock, a configuration reminiscent of Earth's then understood composition before Lehmann's inner core discovery [65].

In ordinary chondrite meteorites, nickel is *always* found alloyed with iron metal; all of the elements heavier than iron and nickel, even combined together, could not comprise a mass as great as the inner core. So what is the composition of the inner core?

In 1940 Birch [66] thought he had the answer. Birch *assumed*, without corroborating evidence, that the inner core is iron metal in the process of solidifying (freezing) from the liquid iron-alloy core (like an ice cube in a glass of ice-water). If Birch were correct, one could determine the temperature at the inner core boundary by measuring the solidification temperature of iron at the respective pressure. That is what Li et al. [61] did in 2020 and which has been done by many since the 1940s, but the basis is a fatally-flawed assumption.

For 39 years Birch and other geoscientists had no reason to believe the inner core composition was other than partially frozen iron (or nickel-iron) metal.

When Birch [66] and others imagined that Earth resembled an ordinary chondrite meteorite, they ignored a different possibility, an *enstatite chondrite*, one of the much less common chondrite meteorites whose matter had formed under oxygen-starving conditions and even contained some minerals not found on Earth's surface. Because of their rarity and seemingly inexplicable oxygen-starved minerals, enstatite chondrites were simply ignored as candidates for Earth's interior composition.

In 1976, Hans E. Suess and I [67] discovered that the oxygen-poor composition of enstatite chondrites' parent matter could be understood as a consequence of their having condensed at high temperatures and high pressures from a gas with the composition of the sun, provided that matter was isolated from further reactions at lower temperatures. In that medium, at higher pressures substances condense at higher temperatures, but the reaction that makes available oxygen is independent of pressure and limits availability of oxygen at higher temperatures.

Because of the oxygen-starvation of enstatite chondrite parent matter, a portion of their elements that love to combine with oxygen, instead of residing entirely in the oxygen-loving silicate-rock portion, occur in part in the iron alloy portion. These elements include calcium, magnesium, silicon, and uranium.

While studying enstatite chondrite meteorites in the 1970s, I realized that, if silicon were present in Earth's core, it would combine with nickel as nickel silicide, which would form a mass at the center almost identical to the mass of the inner core.

Then in 1979, I published a contradiction [4] to the 39 year old inner core idea (Figure 2.4).

> Proc. R. Soc. Lond. A **368**, 495-500 (1979)
> Printed in Great Britain
>
> ## The nickel silicide inner core of the Earth
>
> BY J. M. HERNDON
>
> Department of Chemistry, University of California, San Diego,
> La Jolla, California 92093, U.S.A.
>
> (Communicated by H. C. Urey, For.Mem.R.S. – Received 27 November 1978
> – Revised 19 April 1979)
>
> From observations of nature the suggestion is made that the inner core of the Earth consists not of nickel–iron metal but of nickel silicide.
>
> Contemporary understanding of the physical state and chemical composition of the interior of the Earth is derived primarily from interpretations of seismological measurements and from inferences drawn from observations of meteorites. Seismological investigations by Oldham (1906), Gutenberg (1914) and others helped to establish the idea that a fluid core extends to approximately one half the radius of the Earth. The existence of a small, apparently solid inner core at the centre of the Earth was recognized by Lehmann (1936) from interpretations of

Figure 2.4. From [4].

Figure 2.5 is the image of a complimentary letter I received from Inge Lehmann in which she expressed interest in the responses of other geophysicists. Now, forty-five years later I review those responses.

```
p.t.Søbakkevej 11
    2840 Holte, Denmark                    August 17, 1979


Dr. J.M.Herndon
Department 6f Chemistry
University of California, San Diego
La Jolla, California 92093


Dear Dr. Herndon,

    Thank you for sending me your very interesting paper:
Earth's nickel silicide inner core.

    I admire the precission of your reasoning based on
available information, and I congratulate you on the highly
important result you have obtained.

    It has been a special pleasure to be informed in advace
of publication. I shall be interested to note the reactions of
other geophysigists.

    With kind regards
                                        Yours sincerely
                                                       Inge Lehmann
```

Figure 2.5. Letter from Inge Lehmann to the author.

While awaiting publication of my nickel silicide inner core paper [4], I imagined that there would be debate and discussion, and worried that geoscientists with well-funded laboratories would pick up the ball and run with it, leaving me in their dust. Instead there was silence. It was as if the paper had never been published. That work was ignored and has been ignored for forty-five years, as evidenced, for example, by Li et al.'s 2020 paper [61]. Moreover, my NASA grant, which had funded the work, was not renewed, for no good reason. I was "excommunicated" and without that grant my university position evaporated.

Science, properly executed, is a logical progression of understanding. One new discovery, if correct, potentially leads to a series of successive discoveries. An incorrect "discovery" leads nowhere, trapping those blind adherents in an intellectual cul-de-sac: That is what happened to the geoscience community as a result of ignoring my 1979 fundamental nickel silicide scientific article.

Even at that time I realized that, if I were correct about the nickel silicide inner core, then most of the current scientific understanding about Earth's origin, composition, and behavior is wrong.

But was I correct? One question to ask is which of the chondrite meteorites have a sufficiently great weight percent of iron alloy to match the weight percent of Earth's iron alloy core. The data, shown in Figure 2.6, leave no doubt that only the *enstatite chondrites*, not *ordinary chondrites*, are sufficiently rich in iron alloy to match Earth. Consequently, the rationale upon which Birch [66] based his inner core interpretation is baseless.

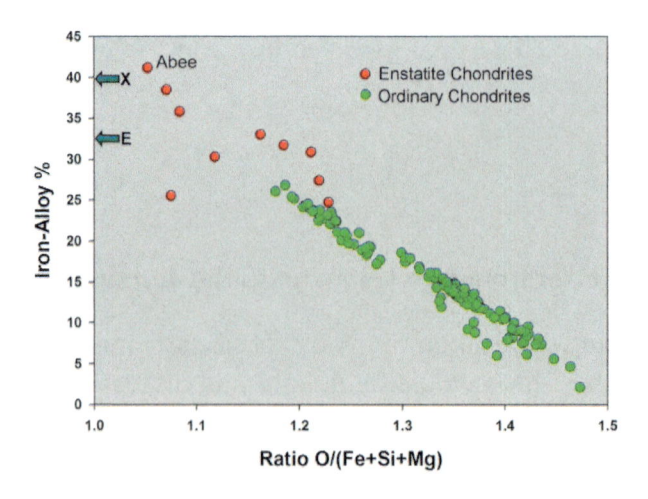

Figure 2.6. Comparison of the mass percent of iron alloy in various chondrite meteorites to that of the Earth as a whole (E) and the endo-Earth (X) (lower mantle plus core [68]).

The composition of Earth's inner core is not an isolated, disconnected entity, but is inextricably related to Earth's origin and composition.

On June 9, 1952 the Abee enstatite chondrite fell to ground in Alberta, Canada [69]. Figure 2.7a shows a nearly complete slice of the roughly basketball-size, 107 kg Abee enstatite chondrite. Abee has been described as an explosion breccia because of its angular fragments[70], but its morphology is quite unique. Peripheries of some of the angular components are shiny, enriched in iron metal that was molten. Figure 2.7b, is a micrograph showing crystals of the major silicate-mineral, enstatite ($MgSiO_3$) embayed (surrounded) by iron which was liquid at a time when the mineral crystal was solid. Figure 2.7c is a micrograph of the iron metal, etched with acid that reveals platelets of *pearlite*, iron carbide, indicative of relatively rapid cooling. M. Lea Rudee and I in 1978 [71] and 1981 [72] published the results of metallurgical experiments that showed during its formation Abee last cooled from 700°C to 25°C in ten hours.

Figure 2.7. (A) Nearly complete slice of the Abee enstatite chondrite. (B) Micrograph showing its enstatite crystals surrounded by previously molten iron metal. (C) Micrograph showing platelets of iron carbide in its metal.

Follow this logical progression which I first considered in 1980 [68]: If the inner core is indeed nickel silicide, then the core must be like the alloy portion of the Abee enstatite chondrite meteorite, which means that Earth's core should be surrounded by a silicate-rock shell like Abee's enstatite silicate ($MgSiO_3$). Multiplying the mass of Earth's core times Abee's silicate to alloy ratio [73] yielded the mass of the silicate shell that must surround the core. I found that the radius of that silicate shell corresponds within 1% to the location of the seismic boundary that separates the lower mantle from the upper mantle [74]. Thus, the ratios of mass for the internal shells of the Earth (inner core, total core, lower mantle) should match those of the Abee enstatite chondrite meteorite, and they do, as shown in Table 1.

Later, I realized that calcium and magnesium, additional elements in the core with high affinities for oxygen, would combine with sulfur to form calcium sulfide (CaS) and magnesium sulfide (MgS), respectively, and float to the top of the core. These components also can be connected with parts of Earth by mass ratios, as shown in Table 1; for details, see [10].

Table 1. Comparison of fundamental Earth mass ratios with corresponding ratios for the Abee enstatite chondrite

Fundamental Earth Ratio	Earth Ratio Value	Abee e.c. Ratio Value
Lower Mantle Mass to Total Core Mass	1.49	1.43
Inner Core Mass to Total Core Mass	0.052	theoretical 0.052 if Ni_3Si 0.057 if Ni_2Si
Inner Core Mass to Lower Mantle + Total Core Mass	0.021	0.021
D" CaS + MgS Mass to Total Core Mass	0.09	.011
ULVZ of D" CaS Mass to Total Core Mass	0.012	0.012

The mass ratio relationships, shown in Table 1, are compelling evidence that the interior 82% of Earth (lower mantle plus core) resembles an enstatite chondrite. Moreover, the calcium sulfide and magnesium sulfide mass ratio relationships solve another problem geoscientists have wrestled with for decades.

Since 1938 seismologists have observed "roughness" or "islands of matter" at the interface between the core and lower mantle [75-77] which geoscientists attempt in various ways to explain as originating from above the core [78-80]. The mass ratio relationships, shown in Table 1, on the contrary, provide compelling evidence that the "islands of matter" at the core-mantle boundary originate from within the core. Moreover, the totality of the relationships in Table 1 clearly indicates that the endo-Earth, core plus lower mantle [68], strongly resembles the Abee enstatite chondrite. A schematic representation of Earth's interior layers consistent with Table 1 is shown in Figure 2.8.

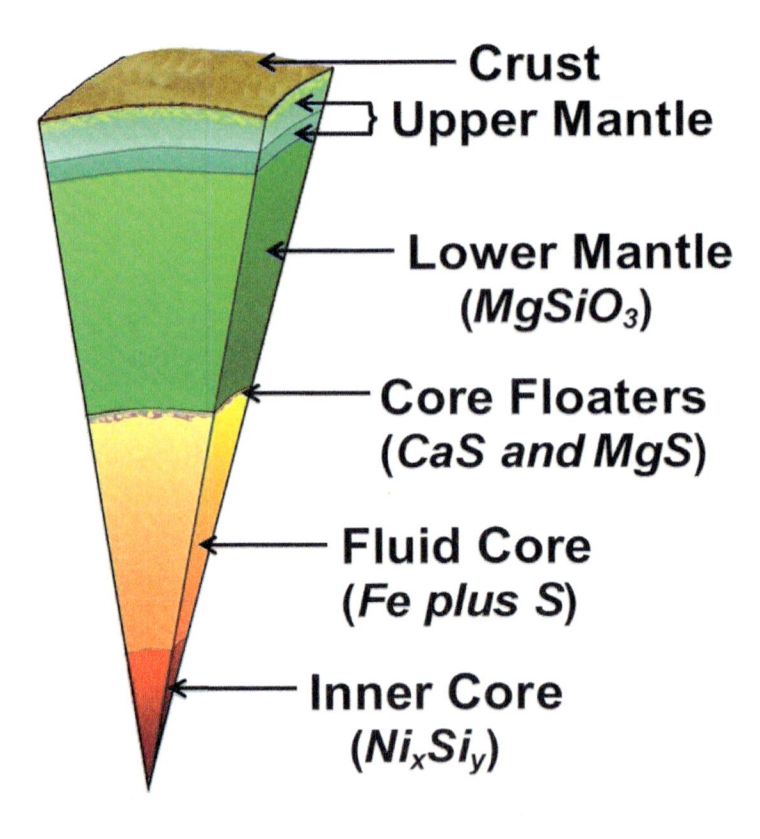

Figure 2.8. Schematic representation of the interior parts of Earth as indicated by the mass ratio relationships shown in Table 1. For details see [81].

In an article published in *Naturwissenschaften* in 1982 [82], I pointed out the importance of determining whether uranium resides in the alloy component of enstatite chondrites. Serendipitously, in 1982 Murrell and Burnett [83] discovered that virtually all of the uranium in the Abee enstatite chondrite resides in its alloy portion. Because Earth's core is virtually identical to the alloy portion of the Abee enstatite chondrite, according to Table 1, one may therefore infer that a very large proportion of Earth's uranium exists in its core, not in its rocky mantle as often assumed by the geoscience community [84].

The next step in my logical progression of understanding was realizing that uranium in Earth's core would settle at the very center of the Earth. In 1993 and in subsequent publications, I applied Fermi's nuclear reactor theory [85] to demonstrate the feasibility of an accumulation of uranium at Earth's center functioning as a nuclear fission reactor, called the *georeactor*, as the energy source for the geomagnetic field [5, 19, 20] (Figure 2.9).

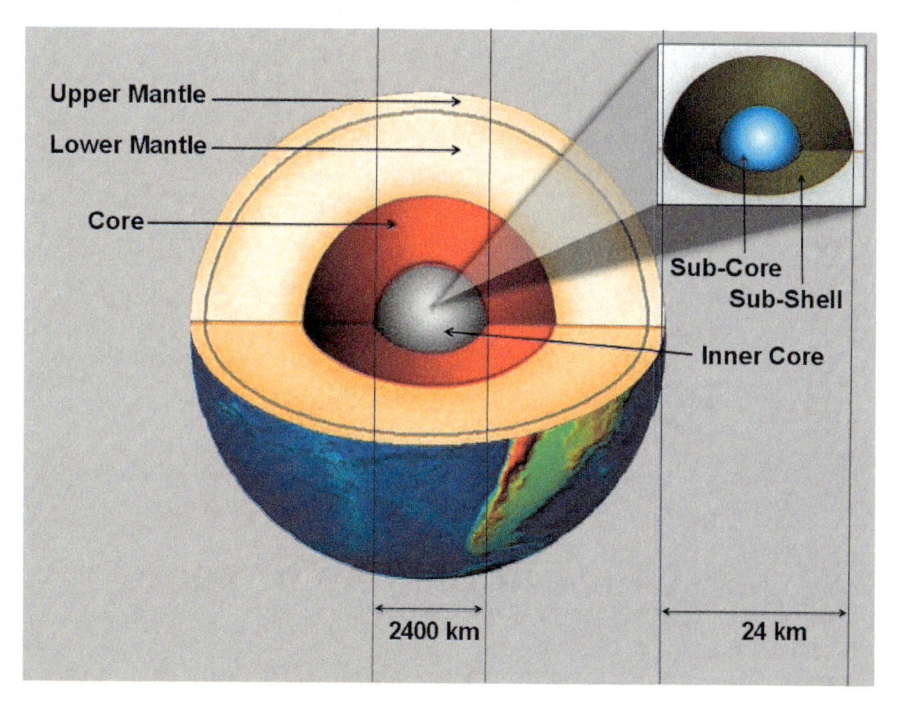

Figure 2.9. Schematic representation of the georeactor, a natural planetocentric nuclear fission reactor.

Fermi's nuclear reactor theory is useful, but does not yield some information, for example, fission products. For decades, Oak Ridge National Laboratory has been developing software to simulate the operation of a variety of nuclear reactors. Dan Hollenbach graciously agreed to modify that software to permit georeactor simulations. The Oak Ridge data confirmed that the georeactor could operate over the lifetime of Earth as a fast neutron breeder reactor and also showed that the fission products include helium-3 and helium-4 in precisely the ratios observed exiting Earth [21, 86].

Figure 2-10 shows georeactor calculated helium-3 to helium-4 ratios relative to atmospheric helium for comparison with their observed ranges in oceanic basalts. The measured basaltic helium-3 to helium-4 ratios provide the first independent, compelling evidence of nuclear georeactor existence.

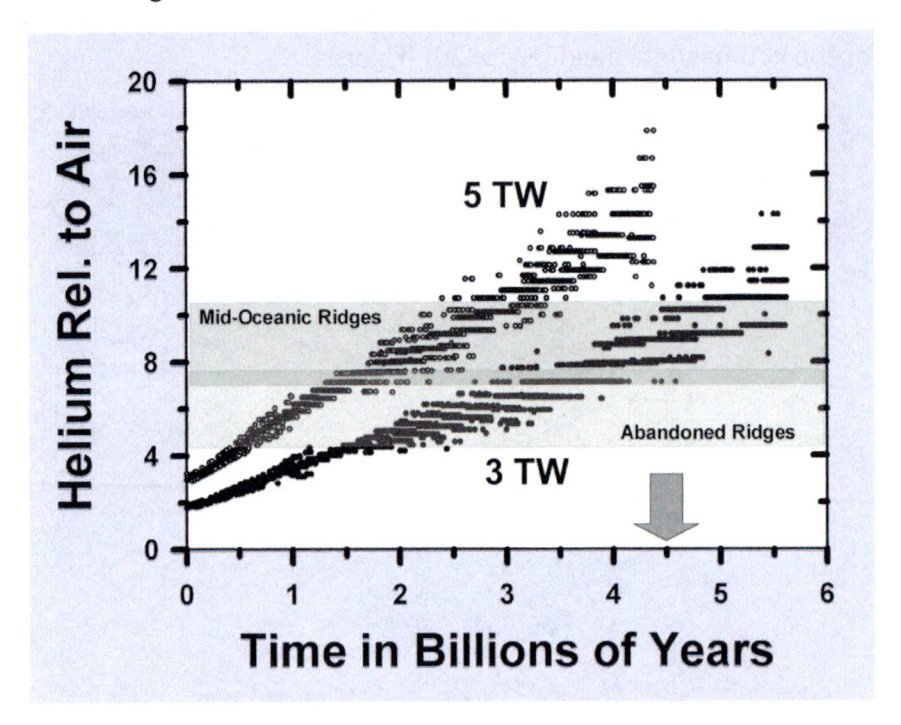

Figure 2-10. Oak Ridge georeactor simulation data calculated at energies of 3 and 5 terawatts compared to measured helium ratios in oceanic basalts. Arrow shows present age of Earth. Data from [6].

In a news article *in Current Science*, Associate Editor K. R. Rao [87] noted that a nuclear reactor at the core of the Earth is *"a solution to the riddles of relative abundances of helium isotopes and to geomagnetic field variability"*. The helium riddle referred to by Rao [87] is this: Since measurements were first made in the 1970s, the helium-3 to helium-4 ratio determined in volcanic basalts typically ranged from 4 to 49 times the same ratio measured in atmospheric helium [88-92]. As there was no deep-Earth mechanism known for producing helium-3 in the requisite quantities, the geoscience community made mantle-mixing computational models based upon the *ad hoc* assumption that primordial trapped helium-3 was mixed with radiogenic helium-4 in just the correct proportion to yield the observed ratios [93-95].

The geoscience community never credited me with the discovery of the origin of deep-Earth helium, but mantle mixing models ceased to be made. Although as late as 2020, some geoscientists began again promoting the same model-nonsense [96, 97].

Note the georeactor helium ratio data shown in Figure 2-10 increase over time. Some helium measurements in basalt from Iceland display high helium ratios, as high as 50 [98]. To me the high helium ratios mean that the georeactor will run out of nuclear fuel at some yet unknown time in the future. In 2002, I submitted a manuscript to the *Proceedings of the National Academy of Sciences* (PNAS) about this. The manuscript was reviewed and was about to be accepted for publication, when suddenly I was advised that PNAS would obtain additional anonymous reviews. I later learned what happened.

The PNAS Editor offered NAS member Don Anderson the opportunity to write a Commentary to accompany my article. Anderson, however, had a conflict of interest, having published a different *ad hoc* idea about the helium [89]. Instead of disclosing his conflict of interest, Anderson convinced the Editor-in-Chief, a biologist, that my paper was deficient and should be further reviewed, of course, by reviewers who owed Anderson for their membership in the National Academy of Sciences (NAS). After two rounds of anonymous reviews by NAS members, in which the three reviewers could make no substantive criticisms, I learned what happened and told the Editor-in-Chief that PNAS integrity had been compromised; the paper was finally published in 2003 [6].

Recently, the Editorship of PNAS devised a different means to suppress articles they do not want to see published. Without even seeking reviews, a member of the PNAS Editorial Board may reject a manuscript simply because the member deems *"it is not of sufficient interest for the readership of PNAS."* That response must surely cause U.S. President Abraham Lincoln to turn over in his grave! President Lincoln chartered the NAS to provide scientific advice to the U. S. Government.

The earlier PNAS article by Hollenbach and me [21] attracted the interest of Brad Lemley, who penned the cover story about my work for the August 2002 issue of *Discover* magazine. Shortly after its publication, I was contacted by an intern at Bell Labs (Lucent Technologies, Inc.). She was planning to give a lunchtime seminar about the georeactor and asked for additional information.

One of the attendees at her talk was R. S. Raghavan, who had published an important article about measuring the elusive, hard to detect antineutrinos to determine radioactive elements in the Earth's interior [99]. Soon after that talk, Raghavan posted on a pre-print server an article entitled *"Detecting a Nuclear Fission Reactor at the Center of the Earth"* [100]. Raghavan [100] showed that the antineutrino spectrum resulting from nuclear fission has a higher energy component than from radioactive decay thus in principle permitting georeactor discrimination.

Despite Raghavan's stellar track record in physics, that article was allegedly rejected for publication in two scientific journals, *Physical Review Letters* and *Physics Letters*. I suspect that it was rejected by physicists and/or geophysicists to cover up the fact that for decades the physics and geophysics communities have been deceiving government science-funding officials, the scientific community, and the public.

Raghavan's article [100], never published but posted on a physics archive, was timely, and stimulated discussions worldwide [101-104]. For example, Russian scientists [105] remarked, *"Herndon's idea about georeactor located at the centre of the Earth, if validated, will open a new era in planetary physics."*

At the time several large-scale antineutrino detectors were under construction or were being considered. The first to reach operational stage was the Kamioka Liquid Scintillator Antineutrino Detector (KamLAND), a joint Japan/American project.

In July 2005, in a paper published in *Nature*, the KamLAND consortium reported the first detection of antineutrinos originating from within the Earth [106]. But what the paper said and what it should have said are two entirely different things. In easy to understand terms, this is what the paper should have said: *In just over two years of taking data, a total of 152 "detector events" were recorded. After subtracting for the background from commercial nuclear reactors and making corrections for contamination, only 20-25 "detector events" were considered to be from antineutrinos originating within the Earth. Within the limitations of the experiment, it is absolutely impossible to ascertain the proportion of those that may have resulted from the radioactive decay of uranium and thorium, or may have been produced from a nuclear fission georeactor at the center of the Earth.*

Instead, what the 87 authors of the KamLAND consortium did was to mislead the scientific community and the general public by wholly and intentionally ignoring the possibility of georeactor-produced antineutrinos. Raghavan's 1998 paper on measuring the global radioactivity in the Earth was cited [99], but not his 2002 paper "*Detecting a Nuclear Fission Reactor at the Center of the Earth*" [100]. And, there was absolutely no reference to any georeactor article.

The KamLAND *Nature* misrepresentation was undergirded by a "News and Views" companion article in the same issue [107] that discussed radioactive decay heat production in the Earth, noting: "*The remaining heat must come from other potential contributors, such as core segregation, inner-core crystallization, accretion energy or extinct radionuclides – for example the gravitational energy gained by metal accumulating at the centre of the Earth, which is converted to thermal energy, and the energy added by impacts during the Earth's initial growth.*" Absolutely no mention was made of georeactor-produced heat, which is on a firmer scientific foundation than some of the "*other potential contributors*" mentioned.

For Japan, the detection of geo-antineutrinos by the KamLAND consortium should have been cause for celebration; instead it was cause for shame. Rather than confronting new and contradictory ideas, American geoscientists have a long and documented record of attempting to prevent their publication and/or simply ignoring them, thereby misleading government research-funding officers, scientists, and the public. In announcing the detection of geo-antineutrinos, Japanese KamLAND scientists, instead of standing tall in integrity, became party to the same American anti-science behavior, and in doing so dishonored themselves and Japan. Curiously, all that was really required in their paper was one carefully worded sentence with appropriate references.

To their credit, after I complained to Japan's Minister for Science and Technology about the 2005 *Nature* misrepresentation, the georeactor was cited in their future publications, although usually just a brief mention among lengthy discussions of models based upon assumptions [108].

Twenty years after my first paper demonstrating the feasibility of a nuclear fission reactor at the center of the Earth as the energy source for the geomagnetic field [19], much development and understanding took place [5, 6, 8, 9, 20, 86]. There had been no published review articles on the georeactor, so I wrote one and submitted it to the Elsevier journal *GeoResJ*. The assigned Editor, a Professor of Geology at the University of Oxford, with no training in the subject of nuclear reactors, rejected the review article, without referee reviews, with a few unwarranted, pejorative remarks. I complained to the University of Oxford's Registrar and Vice-Chancellor, but to no avail.

When academic transgressions occur, such as unwarranted rejection of my papers by faculty members, I frequently file appeals to university presidents and sometimes to regents. But these appeals are *never* successful. Grants are typically made to universities, not to their faculty members. University officers, signatories for government grants, should be those in authority and should maintain integrity. But in my experience, that is *never* the case.

Here is a contrast in intellectual integrity: Following the *GeoResJ* rejection, I submitted the manuscript to *Current Science*, which since 1932 has been published in association with the Indian Academy of Sciences. In that instance, the Editor sent the manuscript to knowledgeable referees who asked for clarification and who asked me to provide additional information, which I did. And it was published [13], entitled "Terracentric Nuclear Fission Georeactor: Background, Basis, Feasibility, Structure, Evidence and Geophysical Implications".

The two currently operational deep-Earth antineutrino detectors, at Kamioka, Japan [108] and at Grand Sasso, Italy [109], to date have not only failed to refute georeactor nuclear fission, but at a 95% confidence level, have measured georeactor energy production of 3.7 and 2.4 terawatts, respectively. Notably, the energy production levels used in the Oak Ridge National Laboratory georeactor calculations, indicated in Figure 2.10, ranged from 3 to 5 terawatts [6]. These antineutrino measurements provide the second independent, compelling evidence of the existence of Earth's nuclear georeactor.

Humanity Imperiled: In 1993 and for fifteen years thereafter, I considered the georeactor as being the energy source that powers the Earth's magnetic field [5, 6, 19-21, 86]. Later, I discovered that there are serious problems with the idea articulated in 1939 [50] for geomagnetism production, and realized that the georeactor could serve both as the energy source and the production mechanism for Earth's magnetic field [13, 16, 62]; the same nuclear fission mechanism could also account for magnetic fields of other planets and large moons [9, 12].

In 1600, Gilbert showed that Earth resembles a giant magnet, rather than the geomagnetism being of extraterrestrial origin as some believed [46]. In 1838, Gauss showed that the source of Earth's magnetism is at or near the center of our planet [47]. Earth's magnetism cannot be generated by a permanent magnet, however, because the iron-alloy core is too hot, being above the Curie temperature at which permanent magnetization disappears. So what produces the geomagnetic field?

In 1855, Faraday published his research which led to an understanding that the motion of electrical charges, i.e. electrical current, produces a magnetic field [48]. So, where might motion exist at or near the center of Earth? In 1939, Elsasser published an idea in the first of several scientific articles [50-52] that is 80 years later still considered to be the scientific basis of the presently popular, but incorrect, explanation of the origin of the geomagnetic field in Earth's core.

Elsasser assumed that motion in the Earth's fluid iron-alloy core was caused by convection and that the electrically conducting, moving fluid acts as a dynamo, producing the geomagnetic field [50-52]. Millions of taxpayer-provided dollars have been wasted on computational models that purport to demonstrate generation of the geomagnetic field by convection in the Earth's fluid core by Elsasser's mechanism. But those models are wrong.

Academic geophysicists rarely consider the most important aspect of science, but one often emphasized by Hans E. Suess [110]: Understanding what is *not* known is more important than knowing what *is* known. Lacking knowledge of any other fluid body at or near the center of the Earth, it is nevertheless wrong to simply assume that the existence of the geomagnetic field connotes convection in the Earth's fluid core. I looked more deeply and discovered that sustained thermal convection in the Earth's fluid core is *physically impossible*.

Thermal convection is a physical process that is easy to visualize. Heat a pot of water on the stovetop and add a few tea leaves. Before the water starts to boil, note that the tea leaves, carried along by the fluid, are in motion, bottom to top and top to bottom. Water heated at the bottom becomes less dense (lighter) and floats to the top as the cooler, more dense (heavier) water at the top sinks to the bottom. This is thermal convection. Just like in the fluid core? No. What is not obvious is that the heat being brought to the top of the water by convection is being lost from the surface. For sustained thermal convection in Earth's fluid core, the heat brought to the top of the core must be promptly removed, but that is not possible as Earth's fluid core is surrounded by an insulating blanket, the rocky mantle [10, 111].

The low loss of heat from the top of the core is not the only reason thermal convection is physically impossible. The weight of the matter above compresses the bottom of the core more than it compresses the top of the core. The very small decrease in density at the bottom of the core, caused by heat, is too little to overcome the greater density increase caused by the weight above [10, 111].

I attempted to publish this important contradiction to a long-standing geoscience misunderstanding in *Physical Review Letters*, which is published by the American Physical Society. My submission was rejected without a scientifically valid reason. I appealed the rejection all the way to the Editor-in-Chief, an NAS member who, after consulting with another unnamed NAS member, rejected my manuscript without a valid scientific basis.

Does the physical impossibility of convection in Earth's core mean that Elsasser is completely wrong? No, it only means that, if the geomagnetic field is produced by dynamo action, as suggested by Elsasser, there must be a different location at or near the Earth's center where thermal convection can continuously move electric charges. And there is, as I described [13, 16, 62], within Earth's central nuclear fission georeactor.

Figure 2.11 is a schematic representation of Earth's georeactor at the center of Earth. In the micro-gravity region near Earth's center, the less-dense nuclear waste, mainly fission and decay products, exist as a liquid or slurry sub-shell above the nuclear georeactor sub-core. Fission-produced heat is transported from the nuclear sub-core by convection to the bottom of the inner core which acts as a heat sink that removes the heat thereby permitting sustained convection. As I have described, this is a self-regulated system that, as necessary for a dynamo, produces electrical charges from radioactive decay. As similarly envisioned by Elsasser in different circumstances, the convective motion coupled with motion from Earth's rotation presumably results in dynamo action that produces the geomagnetic field [13, 16, 62].

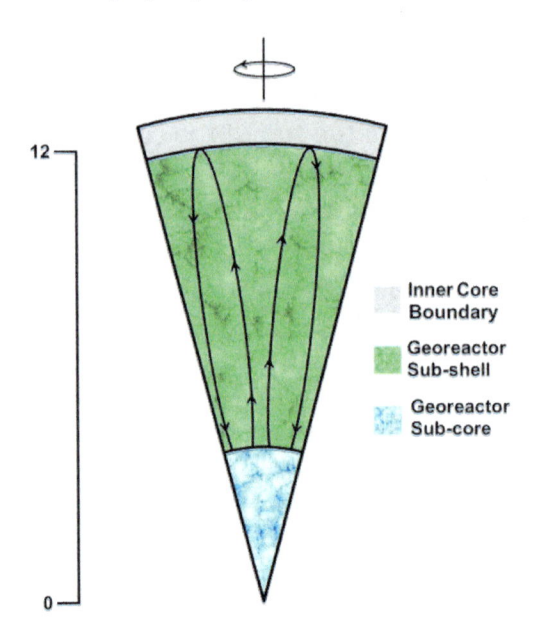

Figure 2.11. Schematic representation of Earth's georeactor, not to scale. Planetary rotation and fluid motions are indicated separately; their resultant motion is not shown. Stable convection is expected with the bottom hotter than the top and with heat removal by the inner core heat sink at the top. Scale in km.

Instead of discovering the true nature of geomagnetism, members of the Deep State geoscience cartel have deceived government officials, the scientific community, and the public. That deception, based upon unrealistic and impossible geomagnetic field generation within Earth's fluid core, has left humanity unaware of the causes and unprepared for the consequences of geomagnetic reversals and/or collapse.

The fluid core is massive – nearly one-third of Earth's mass. The georeactor mass, by comparison, is only about one ten-millionth the mass of the core. That means disruption in georeactor sub-shell convection, which causes geomagnetic reversals and/or collapse, can occur quite quickly [16].

Geomagnetic reversals and/or collapse, caused by disruption of

georeactor sub-shell convection, can potentially occur for the following reasons:
- Massive trauma to Earth, for example, by asteroid impact
- Super-intense solar corona mass ejection
- Anthropogenic disruption of geomagnetic field, for example, by electromagnetic pulse weapons or ionosphere heaters
- Georeactor nuclear fuel burn-up

Williams described some of the consequences to humanity and to our infrastructure that might be expected during geomagnetic reversals and/or collapse [60]: *"Widespread communications disruptions, GPS blackouts, satellite failures, loss of electrical power, loss of electric-transmission control, electrical equipment damage, fires, electrocution, environmental degradation, refrigeration disruptions, food shortages, starvation and concomitant anarchy, potable water shortages, financial systems shut-down, fuel delivery disruptions, loss of ozone and increased skin cancers, cardiac deaths, and dementia. This list is not exhaustive. It is likely that a geomagnetic field collapse would cause much hardship and suffering, and potentially reverse more than two centuries of technological infrastructure development."*

In 2020, I submitted a manuscript to the *Proceedings of the Royal Society of London* in which I described a fundamentally new scientific basis for understanding that, in addition to the widespread adverse consequences described by Williams [60] and quoted above, geomagnetic reversals and/or collapse potentially might also cause major geophysical disasters, for example, triggering super-volcano eruptions. Not unexpectedly, that submission was rejected on the basis of anonymous, false and pejorative reviews devoid of scientific substance. I appealed the Editor's rejection decision, who, instead of appointing an Adjudicating Editor, rejected my appeal himself. I then appealed to the Editor-in-Chief whose background includes space weather, but he refused to consider my appeal.

In 1978, 1979, 1980, and 1994 I published important scientific advances in the *Proceedings of the Royal Society of London* [4, 20, 68, 112]. But publication in that journal appears to be no longer possible. Deep State science corruption seems to have permeated the Royal Society, as it has in other National Academies, and with many scientific publishers, but, fortunately, not all. Noble ideals, such as freedom, truth, and concern for humanity, are not easy to kill, despite concerted efforts to do so.

Government leaders and educators depend upon scientists to describe truthfully and to the best of their knowledge the way Earth's processes work, and to warn of natural and anthropogenic dangers to the environment and to Earth's biota. One hundred years ago, when there was essentially no government support for science, scientists themselves maintained scientific integrity. But after World War II, the glut of government support for science, coupled with flawed funding-administration procedures, progressively led to the Deep State corruption of science [113, 114].

About a decade before I earned the Ph.D. degree in nuclear chemistry in 1974, changes had begun to take place in the physical sciences. Instead of making discoveries, logically and causally connected to the properties of matter and radiation, scientists began making phenomenological models that purport to describe empirical relationships of phenomena to each other [115]. Such models are typically based upon *ad hoc* assumptions, employ computer-based computations and parameters pre-selected to yield an *a priori* desired result. Although a model may appear to emulate some aspect of nature, there is no certainty that nature behaves in the manner modeled. As noted by Box [35], all models are wrong, but a few are useful, for example, models that might predict the path of hurricanes.

When a new idea or observation arises in science, scientists should attempt to refute it. If unable to refute it, scientists, should cite it in subsequent publications on the subject. That is how science progresses, and to be reliable, must progress.

Over the past 45 years, I have made fundamental scientific discoveries that have yielded several paradigm shifts in geoscience. But these advances have been systematically ignored, and at times suppressed, by government-funded scientists functioning in the manner of cartel members. Here I describe a fundamentally new indivisible paradigm that explains Earth's origin and behavior in logical, causally related ways grounded in the fundamental properties of matter and radiation. The advances documented here stand as a reference by which to compare and evaluate the model-nonsense that has been published for decades by government-funded scientists.

Geophysical Cognitive Dissonance: The apparent "fit" of transoceanic continental coastlines (South America with West Africa; North America with West Europe), the matching of *Mesosaurus* fossils in Brazil and those in Ghana, and the matching of sediments, including coal field strata deposited in the Carboniferous, in both Europe and North America, led Wegener [116] in 1912 to conclude that these continents were joined about 330 million years ago in the super-continent, Pangaea, which broke apart and drifted through the surrounding ocean as continents and islands, reaching their present locations in recent times. Fifty years later, upon discovery of ocean-floor magnetic striations, geoscientists recast *continental drift* into *plate tectonics theory*, based upon the idea that continental "plates" move about Earth' surface riding atop putative mantle convection cells [117] – a physical impossibility because the density increase, caused by compression due to the weight above, is too great to be overcome by thermal expansion [10, 41].

In 1933, between the beginnings of *continental drift* and *plate tectonics theory*, Hilgenberg [118] had a different idea. He imagined that at some time in the past, Earth was smaller, its surface completely covered with continental matter, that then expanded, resulting in continent separation. *Earth expansion theory* had its adherents [119], but there were problems. Vast amounts of energy are required for planetary expansion [120, 121]; further, most of Earth's ocean floors are no older than 200 million years. In 1982, Scheidegger [21] stated: *"If expansion on the postulated scale occurred at all, a completely unknown energy source must be found"*. In 1993 and 2005, I discovered two unknown energy sources, georeactor nuclear fission energy [5, 6, 13, 16, 19-22, 24] and the stored energy of protoplanetary compression [7, 12, 122]. With this knowledge, I set forth *Whole-Earth Decompression Dynamics* [7, 10, 123], and resolved geophysical cognitive dissonance by the still dominant theories of plate tectonics and continental drift.

Whole-Earth Decompression Dynamics (WEDD): The primary energy source for geodynamics and supplemental nuclear fission energy are both direct consequences of our planet's protoplanetary formation as a Jupiter-like gas giant.

Primordial condensation at high pressures and high temperatures resulted in oxygen-starved elemental matter within the Earth, including uranium concentrating at Earth's center and functioning as a nuclear fission reactor [5, 6, 8, 13, 16, 19-22, 24, 86].

Earth's complete primordial condensation and aggregation resulted in the formation of a gas giant planet whose rocky interior was surrounded by 300 Earth-masses of ices and gases, a planet similar to Jupiter. At the center, the rocky planetary interior with its fluid core was compressed, by the weight of these ices and gases, to about two-thirds Earth's present diameter.

When the sun entered its T-Tauri stage, presumably during ignition of its thermonuclear fusion reactions, the gases and ices were stripped from the Earth, leaving behind a compressed rocky planet with a contiguous crust devoid of ocean basins [7, 36, 122, 124] (Figure 2.12).

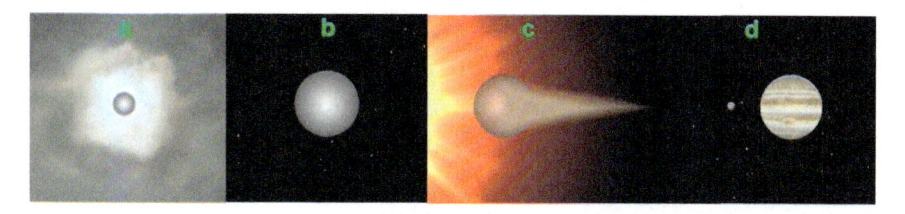

Figure 2.12. *Whole-Earth Decompression Dynamics* **formation of Earth. From left to right, same scale: a) Earth condensing at the center of its giant gaseous protoplanet; b) Earth, a fully condensed gas-giant planet; c) Earth's primordial gases being stripped away by the sun's T-Tauri solar eruptions; d) Earth at the onset of the Hadean eon, compressed to two-thirds of its present diameter showing Jupiter for size comparison.**

During protoplanetary compression by about 300 Earth-masses of ices and gases, the heat of compression was lost. Following removal of the great weight of ices and gases by the young sun's T-Tauri eruptions, Earth, at the onset of the Hadean eon, was compressed to about two-thirds of its present diameter, completely surrounded by a rigid (continental) crust without ocean basins, and containing a vast energy source, the stored protoplanetary energy of compression.

What was Earth like at this point? Its core had already formed; in fact, the core was the first part of Earth to form. The crust and perhaps into the upper mantle was initially quite cold having formed just before the Sun ignited and began stripping away 300 Earth-masses of primordial gases by the super-intense solar wind, which may have cooled the crust even more. There must have been intense bombardment by meteorites and comets in the final stages of Earth formation, which emplaced iron and iron-loving elements, like nickel, in the upper mantle and in the crust.

After the primordial gases and ices had been stripped from Earth's rocky surface, and the violent T-Tauri phase had ended, water brought to Earth's surface by comets began to collect; perhaps the water was brought by the small comets described by Frank [125, 126], which he asserts continue to bring water to Earth today. Volcanic eruptions may have contributed water as well. In the absence of deep ocean basins, inland seas eventually covered much of Earth's surface. Oceanic features, such as pillow basalts from underwater volcanic eruptions and banded ironstone deposits (Figure 2.13), consequently are found within continents [127].

Figure 2.13. Banded ironstone from North America, formed 2.1 billion years ago, presumably during a transition period from less available oxygen to more available oxygen. Photo courtesy of André Karwath.

Meanwhile, deep within the Earth, pressures were building. Occasionally there would be a "blow out". Pressure would force a column of matter from a depth of about 150 km to puncture a narrow hole a few meters in diameter through all of the overlying rock and explode at the surface in a funnel shape as wide as 200 meters [128]. The eruptions of these diamond-bearing kimberlite pipes, however, were just sporadic events. Major catastrophic geological violence would occur again and again, as whole-Earth decompression split the continental crust, created new ocean basins, produced mountain ranges characterized by folding, and caused widespread species extinction.

Earth's behavior, described by *Whole-Earth Decompression*

Dynamics [7, 10, 122, 123, 129], is the basis for virtually all surface geology and geodynamics.

Even though it possessed the two powerful energy sources needed for decompression, the stored energy of protoplanetary compression and nuclear fission energy, whole-Earth decompression was impeded by several factors. For decompression to progress, heat must be added to replace the lost heat of compression. Unless heat is added, decompression would cause cooling and impede decompression. The relative rate of decompression is also a function of *rheology*, the manner by which matter responds to deformation. Furthermore, much greater pressure is required to initiate cracks than to subsequently extend those cracks in the rigid crust.

Nuclear fission energy and the decay energy of radioactive nuclides within the Earth provide sufficient heat to replace the lost heat of protoplanetary compression. As decompression proceeds, Earth's surface responds in two fundamental ways, by increasing surface area and by altering surface curvature.

As described by *Whole-Earth Decompression Dynamics* [7], during whole-Earth decompression, as Earth's volume increases, its surface area increases by the formation of decompression cracks. *Primary decompression cracks* with underlying heat sources extrude hot basalt-rock, which flows by gravitational creep until it falls into and infills *secondary decompression cracks* that lack heat sources.

The chains of volcanoes that form the mid-ocean ridge system, encircling Earth's surface like stitching on a baseball (Figure 2.14), represent a major system of primary decompression cracks. Basalt extruded from these volcanoes forms new seafloor, and flows by gravitational creep across the ocean basins until it falls into and infills secondary decompression cracks that are often located on continental margins. Prominent examples of secondary decompression cracks include the circum-Pacific trenches (Figure 2.15).

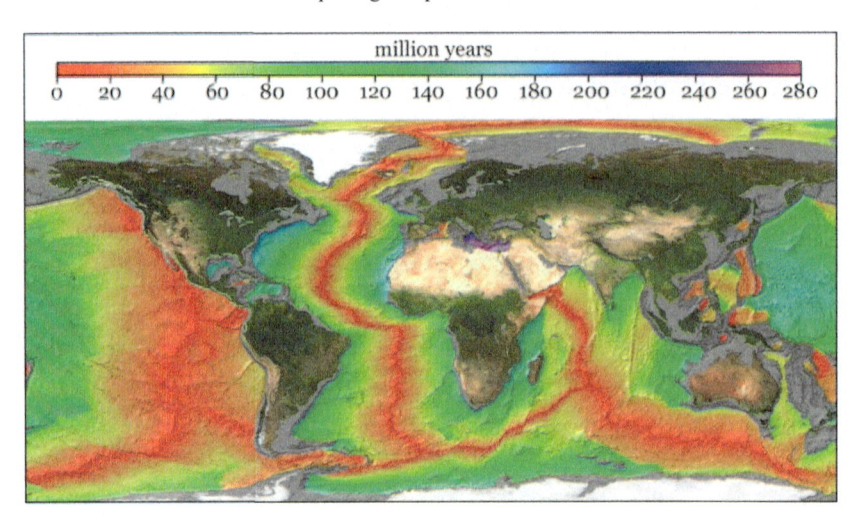

Figure 2.14. U. S. National Oceanic and Atmospheric Administration image showing the ages of ocean floor basalt extruded from volcanoes of the mid-ocean ridge system.

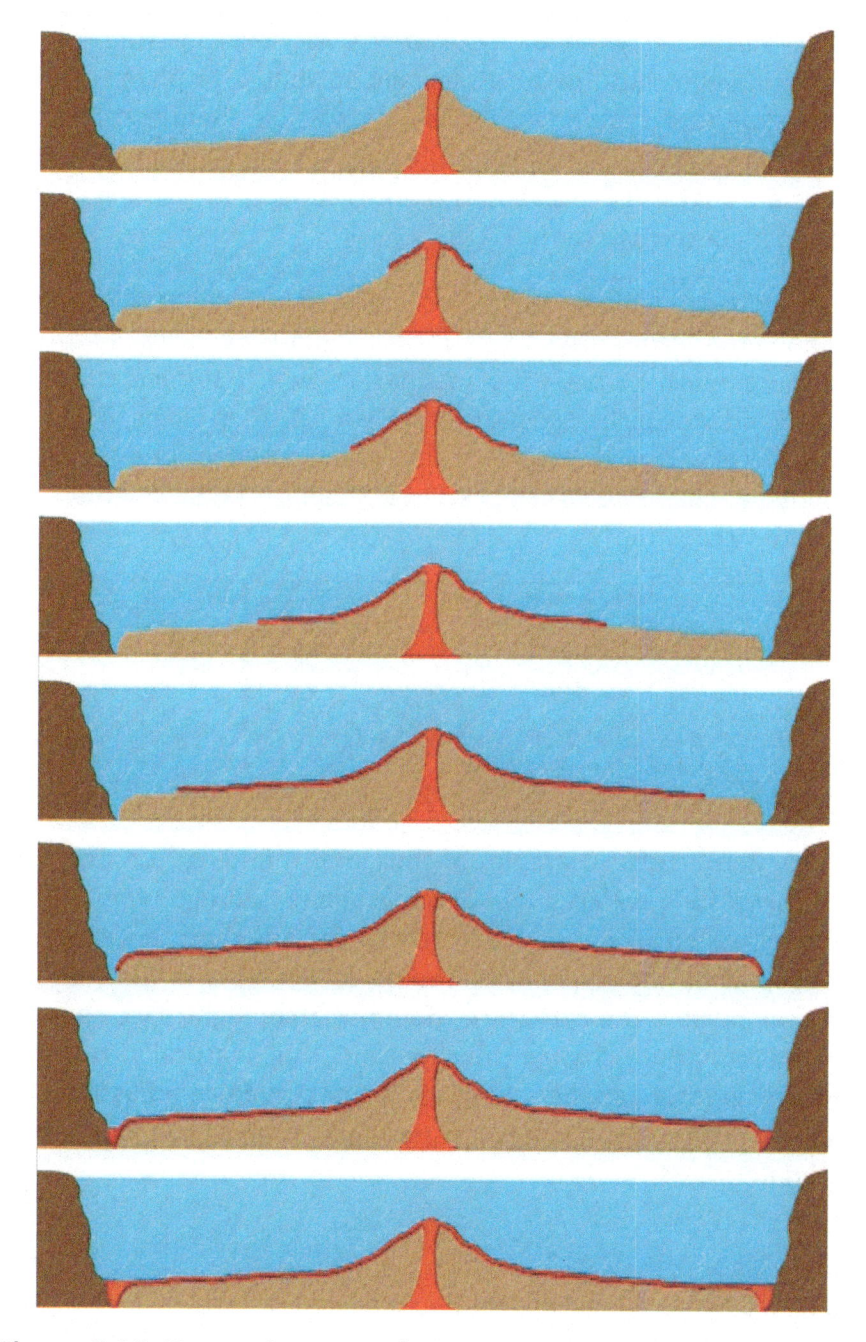

Figure 2.15. Top to bottom schematic image of the Whole-Earth Decompression Dynamics process of increasing Earth's surface area by the formation of ocean basins. Courtesy of Seyo Cizmic.

Whole-Earth Decompression Dynamics [7] explains the myriad submarine geological features, usually attributed to plate tectonics theory, without requiring physically impossible mantle convection [10]. Plus, *Whole-Earth Decompression Dynamics* [7] explains oceanic troughs, inexplicable in plate tectonics, as partially-infilled secondary decompression cracks.

As described by *Whole-Earth Decompression Dynamics* [7], during whole-Earth decompression, as Earth's volume increases, its surface curvature must change. The manner by which surface curvature alteration takes place, illustrated in Figure 2.16, explains, in logical, causally related ways, major Earth geological features, including mountain chains characterized by folding [129], fjords, and submarine canyons [14].

Figure 2.16. Left: Example of mountain folding; Center: The necessity for surface curvature change during whole-Earth decompression. The un-decompressed Earth is represented by the orange, while the larger, decompressed Earth, is represented by the melon. Note the curvatures do not match; Right: Two causally-related curvature-change mechanisms that naturally result in surface curvature change, namely, major curvature adjustment by folded-over tucks, minor curvature adjustment by continental-perimeter tears.

WEDD: Origin of Fold-Mountain Ranges: The origin of mountain chains characterized by folding (Figure 2.17), among Earth's most conspicuous geological features, have not previously been correctly explained, although attempts were made by Dana [130], La Conte [131], Suess [3], Kossmat [132], and others.

Figure 2.17. Mount Everest in the Himalayan fold-mountain chain.

The origin of mountain chains characterized by folding is a natural consequence of *Whole-Earth Decompression Dynamics* [129]. Increases in planetary volume result in excess surface matter within continental perimeters that formed when Earth's volume was smaller. As illustrated in Figure 3.9, gravity causes the excess continental surface matter to buckle, fall over, and break, thus forming mountain ranges characterized by folding [129]. To a lesser extent, the excess continental surface matter causes decompression-stress-tears around continental edges resulting in the formation of fjords (long, deep, narrow channels; see Figure 2.18) as well as submarine canyons [14].

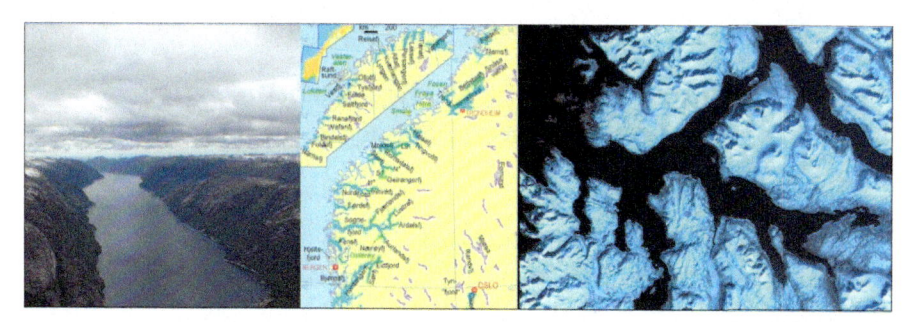

Figure 2.18. (left to right) Photo of Lysefjord, Norway, courtesy of Snorre; Norwegian map showing fjords; Satellite photo of fjords in northern Norway.

Fictitious Super-Continent Cycles: When individual scientists attempt to describe natural phenomena, events, or processes within the binding framework of a problematic paradigm, the explanations they proffer are generally more complex, if not physically impossible, than subsequent, corresponding explanations made within a newer, more-correct paradigm. In plate tectonics theory, mountain formation is thought to be caused by continent collisions [40], since the plates are assumed to move around the globe riding atop so-called mantle convection cells that defy the laws of physics [10]. Within that belief, discoveries of mountain chains older than the assumed formation of Pangaea necessitated the invention of fictitious supercontinent cycles [41], as illustrated in Figure 2.19.

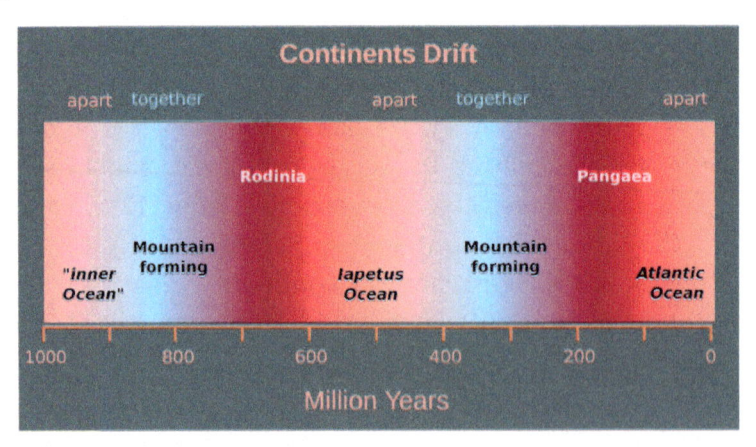

Figure 2.19. Illustration showing the fictional plate tectonics idea of supercontinent cycles. Courtesy of Hannes Grobe. Duplicate from Chapter 1.

A similar problem in plate tectonics arises from rock-magnetism measurements. False rock-magnetism paleolatitude determinations led to the belief that rocks in one location (for example, Vancouver Island, Canada) were thought to have acquired their magnetism in a different location (Baja California, Mexico) [133]. The problem with paleolatitude magnetic measurements, as I discovered [134], is that they are based upon the false assumption that Earth's diameter has not changed over time.

WEDD Heat Transport: The stored energy of protoplanetary compression, as described above, provides the energy for whole-Earth decompression, but requires some additional energy to replace the lost heat of protoplanetary compression. Otherwise, whole-Earth decompression would cool Earth's interior. There is, however, one consequence of whole-Earth decompression that emplaces heat at the base of the crust and produces the *geothermal gradient* within Earth's crust. I call that phenomenon *mantle decompression thermal tsunami* [123].

Earth's matter is layered gravitationally on the basis of density. Earth-decompression, beginning as deep as the bottom of the mantle, will propagate upward through progressively less-dense matter, like a tsunami, until it reaches the rigid crust where compression takes place, producing heat due to compression. Mantle decompression thermal tsunami heats the base of the crust and is the reason that the temperature in the crust increases with depth, constituting the geothermal gradient.

Earth's central nuclear-fission georeactor [5, 6, 13, 16, 19-22, 24] powers and produces its geomagnetic field and aids whole-Earth decompression by providing heat to replace the lost heat of protoplanetary compression. Georeactor heat also channels from Earth's core to its surface [10]. Among its fission products, the georeactor produces helium-3 and helium-4 which serve as tracers that identify georeactor heat channeled to Earth's surface [10] (Figure 2.20). As the uranium fuel is consumed in Earth's Terracentric nuclear reactor, the *helium-3 to helium-4 ratio*, relative to air, as shown in Figure 2.20, increases over time. Helium ratio values of 10 or higher are indicative of recently-produced georeactor helium.

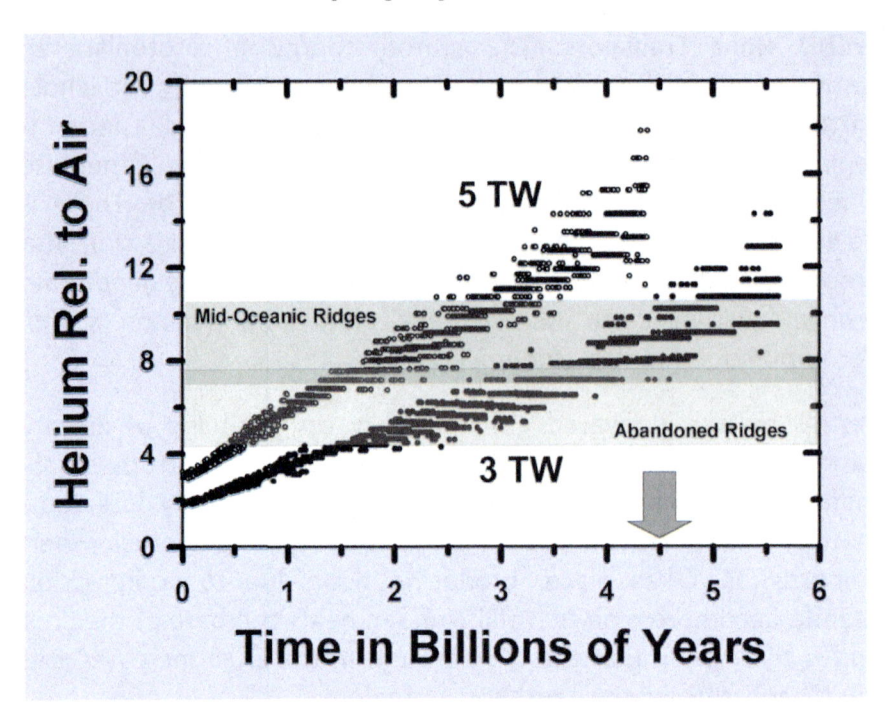

Figure 2.20. Oak Ridge National Laboratory georeactor simulation data calculated at energies of 3 and 5 terawatts compared to measured helium ratios, relative to air, in oceanic basalts. Arrow shows present age of Earth. Data from [6]. Duplicate of Figure 2.10.

Sometimes called mantle plumes, thermal structures or heat channels beneath Iceland and the Hawaiian islands have been seismically imaged as extending all the way to the top of Earth's fluid core [135, 136]. Basalt that erupted at these two locations contains traces of helium with the high-ratio signature of georeactor-produced helium [137]. The heat channels provide paths for the very light, unreactive, very mobile helium to reach Earth's surface [10]. The high-ratio helium is indicative of accompanying heat produced by georeactor nuclear fission chain reactions.

Major basalt floods, containing the high-ratio signature of georeactor-produced helium, occurred in the geologic past, for example, 250 million years ago in Siberia (Siberian Traps) [138] and 65 million years ago in India (Deccan Traps) [139].

Currently, basalt extruded by volcanoes along the East African Rift System [140] and in Yellowstone (USA) [94, 141] contain the high-ratio signature of georeactor-produced helium. The Yellowstone measurements, which indicate that Yellowstone's heat source is the nuclear-fission georeactor, are of serious concern, because Yellowstone is believed to be a potential super-volcano [142-145]. Natural or anthropogenic disruption of the geomagnetic field might trigger eruption of that super-volcano [16, 22, 24].

WEDD: Petroleum Origination: The basis for virtually all surface geology, as described by *Whole-Earth Decompression Dynamics* [7], is that as Earth's volume increases during whole-Earth decompression, its surface area increases by the formation and infilling of decompression cracks, and its surface curvature changes mainly by the formation of mountains characterized by folding [129].

Splitting Earth's continental crust has been a progressive series of events over geologic time, for example, by the mid-ocean ridge system shown in Figure 2.14. That fundamental crust-splitting process is still taking place, for example, in the East African Rift System (Figure 2.21).

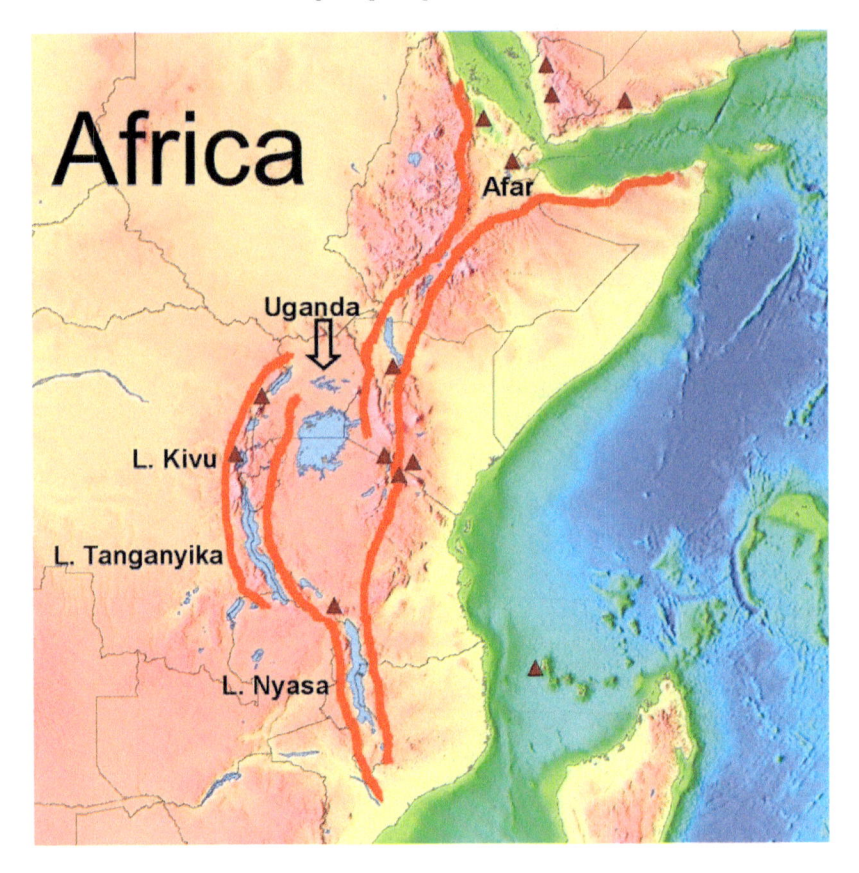

Figure 2.21. East African Rift System indicated in red. Triangles show areas of volcanic activity. From [146].

In 2016, in the *Journal of Petroleum Exploration and Production Technology*, I published an article entitled *"New Concept on the Origin of Petroleum and Natural Gas Deposits"* [147]. That article built upon and extended my two other articles [146, 148] that described the *Whole-Earth Decompression Dynamics* basis for the origination of petroleum and natural gas deposits.

In plate tectonics, the term "rift" refers to the idea that two plates are being pulled apart. In *Whole-Earth Decompression Dynamics*, the geological terms "rift" or "rift zone" pertain to Earth's crust being cracked as a consequence of Earth's volume increasing, which potentially allows mantle gases and organics to escape or be trapped in surface strata. Viewed in this context, it became evident that many, if not most, of the world's great petroleum and natural gas fields are associated with zones where major whole-Earth decompression splitting of the continental crust has taken place, including at the continental margins [147].

Petroleum and natural gas exploration and production are currently underway along the East African Rift System (Figure 2.21), the Rio Grande Rift System in the U.S.A. and in rift systems, basins, and along continental margins that were formerly whole-Earth decompression cracks or failed decompression cracks all over the planet. The West Siberian Basin, host to one of the world's greatest petroleum and natural gas deposits, is the site of a failed whole-Earth decompression crack referred to as the *Siberian Traps*, where massive basalt floods occurred 250 million years ago (Figure 2.22).

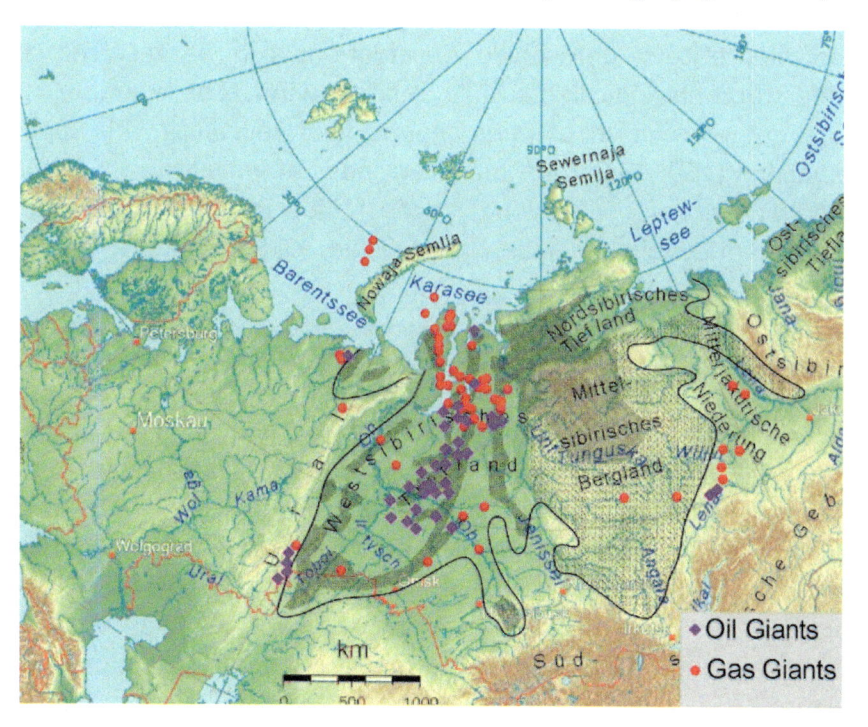

Figure 2.22. Map showing the extent of the *Siberian Traps*, with circles showing major gas fields and diamonds showing major oil fields. From [147].

WEDD: Variations in Sea Levels: In attempting to understand the complex, highly incomplete geological record, much confusion has arisen from interpretations based upon an incorrect paradigm. For example, in the unchanging global-dimension of plate tectonics, the supercontinent Pangaea is thought to be surrounded by ocean. In that view, Pangaea-fragmentation shifted land and ocean volumes around without producing any major change in sea level. The only mechanism envisioned in that paradigm for a rapid, major lowering or raising of sea level was the onset or ending of an ice age, when a large volume of ocean water was sequestered or released as polar and glacial ice [149].

The geodynamics and geology of Earth are intrinsically related through my indivisible geoscience paradigm, *Whole-Earth Decompression Dynamics*. Ultimately, myriads of seemingly complex and theoretically unresolved observations can be resolved and understood in logical, causally related ways. For example, the apparent correlation of geomagnetic field reversals with species extinction [150, 151], with major episodes of volcanism [152, 153], and with drastic sea-level changes [154], is understandable as geomagnetic field collapse, in principle, can lead to a spike in georeactor output energy, and thus possibly trigger a decompression spike manifest, for example, by volcanism, earthquakes, continent splitting, species extinction, and more [16, 22, 24].

The progressive splitting of continental crust and concomitant opening of ocean basins necessarily causes lowering of sea levels, which over time is compensated by new ocean water additions. Continent fragmentation thus exposes sea water to non-oxidized minerals, such as pyrite and arsenopyrite, that can acidify and toxify sea water, and potentially lead to massive species extinctions [155] (Figure 2.23).

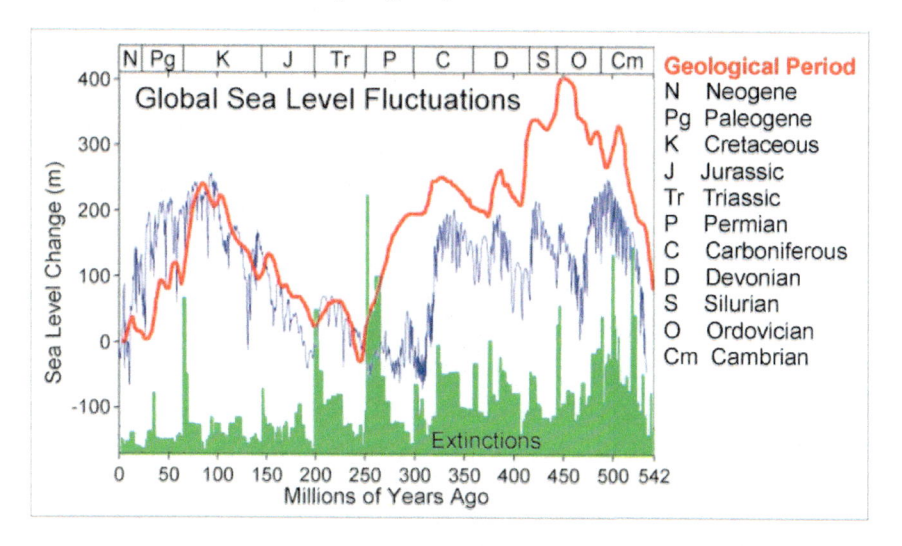

Figure 2.23. Spikes in seawater levels (red and blue) appear to correlate with spikes in species genus extinction intensity (green), and they correlate as well with boundaries of major divisions of geological time, abbreviated at top of graph. For details and data, see [156-163].

Evidence from the geological past is incomplete, but with *Whole-Earth Decompression Dynamics*, the confusion inherent to previous scientifically convoluted explanations for fundamental geological phenomena can be clarified and united with the hopeful result that geoscientists can begin afresh to attain an understanding of Earth's history that is securely anchored to the known properties of matter and radiation.

WEDD Recapitulation: I have described a fundamentally new, indivisible paradigm that recognizes Earth's early formation as a Jupiter-like gas giant and makes it possible to derive virtually all of the geological observations and geodynamic behavior of our planet, including two previously unanticipated powerful energy sources whose absence otherwise raises insuperable dilemmas. Earth's interior condensed from primordial matter at high pressures and high temperatures, with Earth's fluid iron alloy core first raining-out at the planet's center.

Primordial condensation at high pressures and high temperatures progressed on the basis of relative volatility with the first condensate being molten iron. The primordial gas at high pressures and high temperatures led to an oxygen-starved condensate, containing in the fluid iron alloy core, portions of Earth's oxygen-loving elements such as uranium, silicon, calcium, and magnesium. Uranium precipitated and settled at the center of Earth where it eventually began functioning as a nuclear fission reactor, producing the geomagnetic field. Silicon precipitated as nickel silicide and formed Earth's inner core. Calcium and magnesium precipitated as sulfides and floated to the top of the core, forming the seismically "rough" matter observed there.

Following condensation of Earth's fluid core, the principal silicate, enstatite ($MgSiO_3$), condensed and formed Earth's lower mantle, followed by the remaining rocky-matter condensate, mixed with in-falling debris, forming the upper mantle and crust.

Primordial condensation continued with the most volatile substances condensing as ices and gases to form a fully condensed gas giant proto-Earth having a mass almost identical to Jupiter.

Subsequently, violent T-Tauri phase solar winds stripped the ices and gases away leaving, at the beginning of the Hadean eon, a rocky planet that had been compressed to about two-thirds of present-day Earth-diameter, and containing within itself the great stored energy of protoplanetary compression.

Earth's subsequent decompression, described by *Whole-Earth Decompression Dynamics*, in logical and causally related ways, accounts for virtually all of Earth's surface geology and geodynamics.

As whole-Earth decompression progresses and as Earth's volume increases, its surface area increases by the formation of decompression cracks. Primary decompression cracks with underlying heat sources extrude basalt-rock, which flows by gravitational creep until it falls into and infills secondary decompression cracks that lack heat sources. This accounts for the separation of the continents and for the topography of Earth's ocean basins.

As whole-Earth decompression progresses and as Earth's volume increases, its surface curvature must change. The manner by which surface curvature adjusts to changes in volume explains, in logical, causally related ways, the formation of mountain chains characterized by folding as well as fjords and submarine canyons.

Whole-Earth Decompression Dynamics explains, more completely and more correctly, observations usually attributed to plate tectonics without requiring physically-impossible mantle convection or fictitious super-continent cycles. In addition, *Whole-Earth Decompression Dynamics* explains geological observations that are inexplicable by plate tectonics, including the geothermal gradient, oceanic troughs, the origin of petroleum and natural gas deposits, and more.

The observations and discoveries cited-herein have been published by the author in the peer-reviewed scientific literature over a period of four decades. They have rarely been cited by Deep State government-funded scientists. The logical, causally-related advances documented here stand as a reference by which to compare and evaluate the phenomenological model-nonsense that has been published by government-funded scientists for decades at taxpayer expense.

3 PLANETARY SCIENCE MISTAKES AND DECEIT

In 1897, Chamberlain [164] set forth a new hypothesis for planetary formation. In 1900, Moulton [165] modified that hypothesis, which became the Chamberlin-Moulton *planetesimal theory of planetary formation* [166] that explained planetary formation by the accumulation of small bodies.

Beginning in 1963, the *planetesimal theory* became the basis of computational models [167-170] which in aggregate became known by adherents as the *standard model of solar system formation* [171-173]. At the time it was incorrectly believed that the Earth resembled an ordinary chondrite meteorite. The model assumed the minerals of an ordinary chondrite condensed from primordial matter, a hot gas of the composition of the sun at very low-pressures (one ten-thousandth the pressure of the air we breathe) [170, 174]. Then, the condensate progressively gathered into larger rocks, boulders, planetesimals, and finally planets [168, 169]. But the gathered condensate was a homogeneous mixture of iron metal and silicate-rock and all planets have iron metal cores. So, without corroborating evidence, to account for planetary cores, the standard model assumed whole-planet melting with a magma ocean that allowed the more dense molten iron metal to drain down to the planet-center [175, 176].

Phenomenological model-makers typically do not adhere to long-standing scientific principles. For example, in a paper published in the *Proceedings of the Royal Society of London* [112], I utilized thermodynamic considerations to show that, under the assumed low-pressure, hot gas composition of the photosphere of the sun, the condensate would be fully oxidized (unlike the minerals found in ordinary chondrites) and would contain no metal for planetary cores. My work was ignored by the model-makers.

Typically, models are composed of layer upon layer of *ad hoc* assumptions, the consequences of which can often lead to absurdities committed in the name of "science." Consider the planets of our solar system shown in Figure 3.1.

Figure 3.1. Upper images showing the relative sizes of the planets in our solar system. Their relative distances are shown in the lower graph.

The inner four planets are *rocky* while the outer four, the *gas-giants*, contain copious amounts of ices and gases. Lacking corroborating evidence, how were these differences explained by the standard model of solar system formation? It was simply assumed that during primordial condensation there was a temperature gradient across the solar system with an assumed *frost line* between Mars and Jupiter. Beyond the frost line temperatures were sufficiently low to permit condensation of ices and gases, but inside the frost line, temperatures were too high for ices and gases to condense, so that only rocky material could condense.

In the late 1990s, astronomers discovered exoplanets orbiting other stars. Some of these exoplanets were gas giants located as close or closer to their stars as Earth is to the sun. How then could they have formed? To explain this anomaly, astrophysicists invented the concept of *planet migration* wherein gas-giant exoplanets were assumed to have formed in the outer regions of their star systems, and then migrated to where they are currently observed [177].

In 2006, I submitted a brief Letter to *Astrophysical Journal Letters* entitled "Evidence Contrary to the Existing Exo-Planet Migration Concept." The evidence I presented was historical, interdisciplinary, and model-independent. That Letter was rejected out of hand [178]. Suppressing publication of evidence that conflicts with a new unchallenged theory thus allowed planet migration theory to become part of official astro-nonsense – not science [179, 180].

The discovery of close-to-star, gas-giant exoplanets should have been an invitation to make new findings and should have caused astrophysicists to ask the question, "What is wrong with this picture?" Had they asked basic questions that probed their problematic assumption, they might have realized the flaws in their models, and made scientific progress [36].

Nature of Earth's Formation: Meteorites that crash to Earth from space can be categorized into groups on the basis of their chemical compositions. Members of one group, called *chondrites*, are special in that their different non-volatile chemical elements have not been appreciably separated from one another since their origin in a great nuclear furnace. They therefore provide useful knowledge about processes in the solar system at the time planets formed [181-183]. There is a complication, however.

There are three sub-groups of *chondrite* meteorites that differ greatly in their mineral components, because their parent matter formed under quite different conditions, which controlled the amount of oxygen available during formation:
- *Ordinary Chondrites*

- *Carbonaceous Chondrites*
- *Enstatite Chondrites*

Taking thermodynamic considerations into account, I determined that the abundant ordinary chondrites could not have formed in the hydrogen-rich environment thought to have prevailed during their primordial condensation [112, 184, 185], but they must have different origins [12].

The rare, primitive, oxygen-rich *carbonaceous chondrites* are devoid of metal [186, 187] and could not have formed planets with iron metal cores.

The matter from which the rare, primitive, oxygen-starved *enstatite chondrites* formed was an enigma until 1976 when Suess and I [67] demonstrated that primordial condensation at high temperatures and high pressures (10-1000 times the pressure of the air we breathe) would lead to the level of oxygen-starvation found in an enstatite chondrite, provided its parent matter was isolated from the gases at lower temperatures.

Protoplanetary Planet Formation: In 1755, Kant [188] set forth a hypothesis on the origin of the sun and planets that was modified by Laplace [189] four decades later. Laplace's nebula hypothesis was the forerunner of the modern protoplanetary theory of planet formation in which planets are thought to form within giant gaseous protoplanets. The protoplanetary theory attracted scientific attention in the 1940s and 1950s [190-192], but was abandoned and ignored by phenomenological model-makers in the early 1960s who favored the planetesimal theory.

In 1944, Eucken [190] published a scientific article entitled *"Physikalisch-chemische Betrachtungen ueber die frueheste Entwicklungsgeschichte der Erde"* [Physico-Chemical Considerations about the Earliest Development History of the Earth]. From thermodynamic considerations, Eucken investigated condensation from a gas of the composition of the outer part of the sun, mostly hydrogen and helium, but containing small amounts of nearly all of the chemical elements, which is thought to resemble the primordial matter from which the planets formed. Eucken showed that the first primordial condensate from a cooling gas of solar composition at high pressures would be molten iron at high temperatures, followed at lower temperatures by silicate minerals, and at still lower temperatures, by gases and ices. In other words, condensing from within a giant gaseous protoplanet, the formation of Earth began with liquid iron metal raining out to form its core, followed by the condensation of minerals to form its mantle.

Thirty-two years later, while investigating condensation of enstatite chondrite parent material, Suess and I [67] *independently* confirmed Eucken's calculations. The next step was to demonstrate that the core and lower mantle of Earth are essentially identical, respectively, to the alloy and silicate portions of an enstatite chondrite. Using ratios of mass, I related parts of the Abee enstatite chondrite with parts of the Earth [10, 68, 82, 193]. These mass-ratio relationships are shown in Table 1, duplicated from Chapter 2. For details, see [10].

Table 1. Comparison of fundamental Earth mass ratios with corresponding ratios for the Abee enstatite chondrite

Fundamental Earth Ratio	Earth Ratio Value	Abee e.c. Ratio Value
Lower Mantle Mass to Total Core Mass	1.49	1.43
Inner Core Mass to Total Core Mass	0.052	theoretical 0.052 if Ni_3Si 0.057 if Ni_2Si
Inner Core Mass to Lower Mantle + Total Core Mass	0.021	0.021
D" CaS + MgS Mass to Total Core Mass	0.09	.011
ULVZ of D" CaS Mass to Total Core Mass	0.012	0.012

Connecting parts of Earth to enstatite chondrite parent matter, and connecting the oxygen-starved parent matter of enstatite chondrites to primordial condensation at high temperatures and high pressures, therefore connects Earth's formation to high-temperature and high-pressure condensation from within a giant gaseous protoplanet that began with liquid iron metal raining out forming the core, followed by condensation of Earth's mantle minerals.

In 2011, NASA's MESSENGER orbiting spacecraft produced important images of features unique to planet Mercury that were inexplicable to NASA scientists. Many of the images revealed *"... an unusual landform on Mercury, characterized by irregular shaped, shallow, rimless depressions, commonly in clusters and in association with high-reflectance material .... and suggests it indicates activity"* [194] (Figure 3.2).

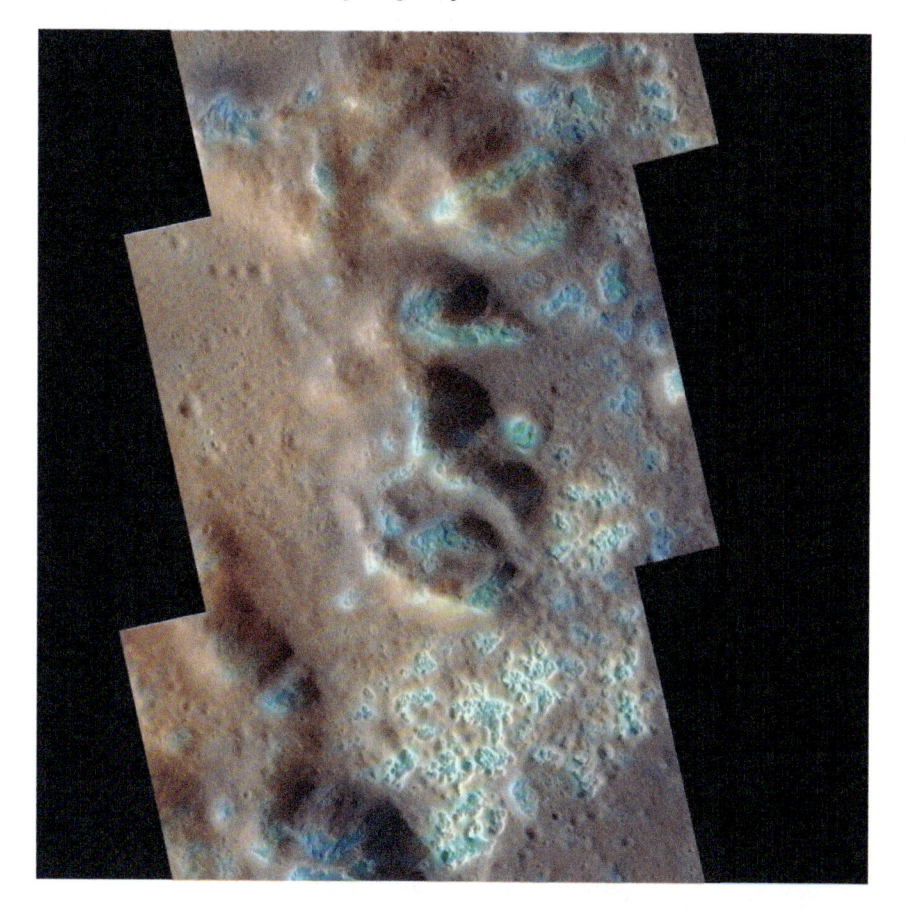

Figure 3.2. NASA MESSENGER image showing pits surrounded by shiny material. These bright shallow depressions appear to have been formed by disgorged volatile matter from within the planet.

In 2012, I published a scientific explanation for the anomalies observed on Mercury's surface [11]. During formation, Mercury's iron core, in condensing and raining-out as a liquid at high pressures and high temperatures from within what was a giant gaseous protoplanet, dissolved a considerable amount of hydrogen, as hydrogen is quite soluble in liquid iron. As Mercury's core solidified, the hydrogen was dispelled and erupted from the surface like hydrogen geysers, forming the surrounding shiny iron metal by turning relatively low reflecting iron sulfide into highly reflecting iron metal.

Figure 3.3 shows the relationship between condensation and dissolved hydrogen. For the indicated hydrogen gas pressures (left vertical axis) and temperatures, the red curve shows the boundary between liquid iron and gaseous iron in an atmosphere like the outer part of the sun. For each temperature/pressure point along the red curve, the amount of hydrogen dissolved in the molten iron, indicated by the blue curve, can be read from the right vertical axis. For reference, the green lines tie together these corresponding points. The hydrogen volume units, at STP (standard temperature and pressure), are equal to the volume of planet Mercury.

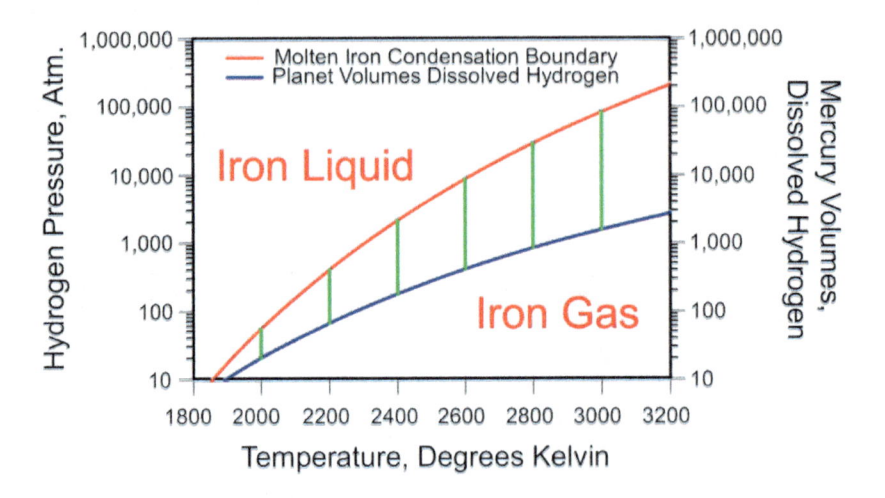

Figure 3.3. By condensing from a giant gaseous protoplanet at pressures above 10 atm., Mercury's core initially was liquid and contained copious amounts of dissolved hydrogen. For details see [11].

Removal of Inner Planet Ices and Gases: If planets formed from giant gaseous protoplanets, as compelling evidence indicates, how were the gases lost from the inner planets (but not the outer planets)?

There is a brief period of violent activity, called the T-Tauri phase, that occurs during the early stages of star formation and is characterized by grand eruptions and super-intense 'solar wind'. A

Hubble Space Telescope image of an erupting binary T-Tauri star is shown in Figure 3.4. The white crescent marks the leading edge of the plume from an observation made five-years earlier.

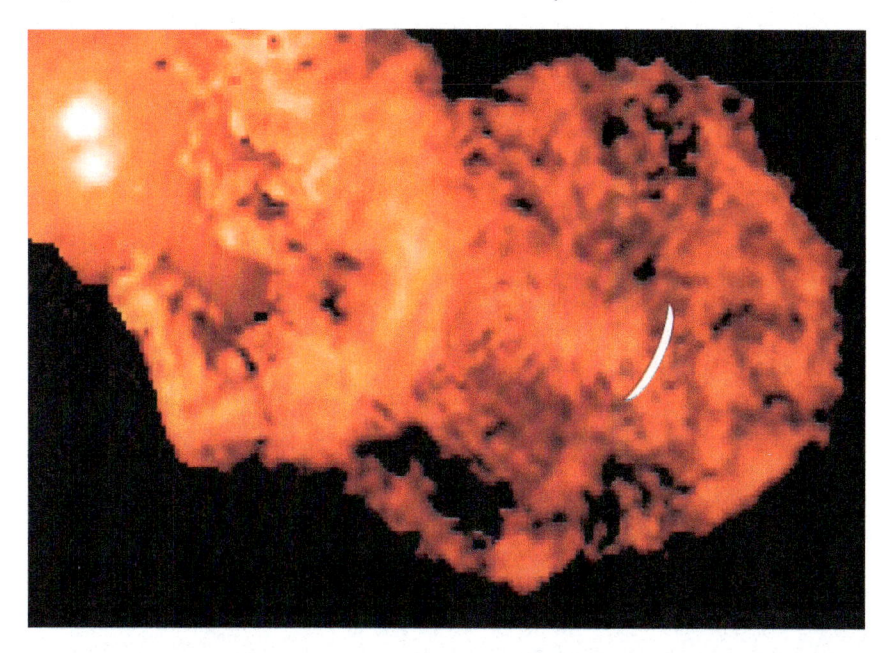

Figure 3.4. Hubble Space Telescope image of a T-Tauri outburst from the binary XZ-Tauri in 2000. The white crescent shows the leading edge of the plume in 1995.

A T-Tauri outburst from our sun, I posit, stripped gas from all the inner planets, and even stripped part of Mercury's incompletely condensed protoplanetary material, and deposited it between Mars and Jupiter where it contributed to the formation of the asteroid belt [12, 185].

Our Moon's Two Different Faces New Paradigm: Much of the temporal biology of life on Earth is regulated and/or influenced by the daily, seasonal, lunar and tidal geophysical cycles [195, 196]. The Moon figures prominently in agriculture [197-199], wildlife behavior [200-203], spiritual matters [204-206], and romance [207-209] (Figure 3.5) and in legend.

As adapted from [210]: *"Chang'E flying to the moon is a beautiful legend in ancient China. Chang'E is a lady graceful of carriage and unparalleled of beauty. After she secretly swallowed the elixir of immortality, she felt herself becoming light. She flew up in spite of herself, drifting and floating in the air, until she reached the palace of the moon. Once on the moon, Chang E became a three-legged toad, as punishment from the Queen Mother [for] going to clasp the moon in the Ninth Heaven."*

Figure 3.5. The inextricable connection between romance and the Moon. Photo by El Salanzo, Unsplash.com.

As noted by Lihua [211]: *"The Moon has fascinated mankind throughout the ages, full of romantic and blue color. By simply viewing with the naked eye, one can discern two major types of terrains on the moon: relatively bright highlands and darker plains.... people like to put the moon into literature particularly poems to express variety of their emotions. The image of the full moon has been endowed with many symbolic meanings over the world, often out of aesthetic need."*

For as long as humans have gazed at the Moon, they have seen the same two-component face, rarely if ever thinking that the opposite face might be substantially different, that the Moon might have two faces. That changed in 1959 when the Soviet Luna 3 spacecraft first photographed the Moon's far side [212].

Lunar maria, the great dark plains, and the lighter highlands are prominent features of the Earth-facing surface of the Moon. The lunar highlands are composed of calcium-aluminum rich anorthositic rocks, whereas the lunar maria are composed of basaltic lava flows [213, 214]. Any attempt to understand the origin of lunar maria must as well account for the fact that maria are nearly absent on the far-side of the Moon (Figure 3.6).

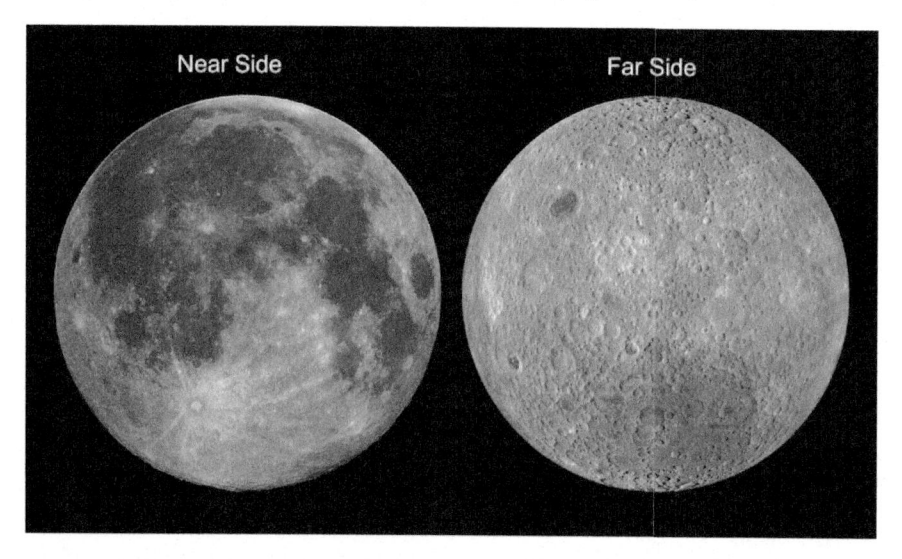

Figure 3.6. NASA image of the near side and far side of the Moon

Scientists have long puzzled about the origin and nature of the Moon [215-217]. In 1974, when I earned the Ph.D. degree, Moon investigations were "all the rage" in the geoscience community. But I abstained, concerned that there were still missing pieces to the puzzle. Then, in 2022, inspiration arrived in a most unexpected way.

Dorion Sagan [218] asked me a question as to the veracity of investigations about the reason for the Moon's facial disparity. Why he wanted to know was both of historical and romantic interest. His question arose while reading love letters from his father, astrophysicist Carl Sagan [219] to his mother, evolutionary biologist

Lynn Margulis [220], a woman who had influenced me in a major way. In the letter, his father, perhaps tauntingly, referred to the "two Marias" (i.e. Moon and Mars) he had been seeing.

Various attempts have been made to explain the origin of lunar maria based solely upon obvious lunar processes, namely, volcanism and impact phenomena [221-223]. These attempts, however, did not lead to any logical, causally related understandings. After being asked the question by Dorion Sagan, the thought occurred to me that the two disparate faces of the Moon might be understood by considering processes on Earth.

Here I review my new paradigm, a quite different, but logical, explanation for the origin of lunar maria [224] derived through analogy with observations of Earth. Also, I describe a way to test the validity of the concept presented, and discuss some potential implications.

Underlying Scientific Background: Phenomena in nature are rarely isolated events, but are usually connected logically and causally by a series of processes. So it is with understanding the nature and composition of the Earth and, as well, the observed disparity of the near side and far side of the Moon (Figure 3.6).

Since the 1940s, numerous geologists and geophysicists have believed that Earth resembles an ordinary chondrite, the most frequently recovered chondrite, which consists of nickel-containing iron metal and silicates. The rare carbonaceous chondrites are so oxidized that iron does not occur as metal. The enstatite chondrites were largely ignored due to their rarity and oxygen-poor mineral composition that was unlike rocks at Earth's surface. For a general review of chondrites, see [225].

My focus on the enstatite chondrites led to the following new concepts and discoveries.
- Solar system formation occurred primarily according to the protoplanetary theory, minimally by the planetesimal theory [12, 122]

- Stored energy of protoplanetary compression as the primary energy driving geodynamics [7, 122, 226]
- Nuclear fission georeactor at Earth's center [5, 6, 13, 19-21]
- Terracentric nuclear fission energy as the secondary energy driving geodynamics [7, 13, 226]
- Basis of heat transport within the Earth [10, 122, 123]
- Earth's magnetic field powered and produced by the Terracentric nuclear fission georeactor [8, 13]
- Whole-Earth Decompression Dynamics, the fundamental basis of geodynamics and geology [7, 26], not requiring physically impossible mantle convection [10], including
    - New concept for the origin of mountains characterized by folding [129]
    - New concept for the origin of fjords and submarine canyons [14]
- Georeactor origin of deep-Earth helium-3 [6]
- Planetocentric nuclear fission reactors as the basis for magnetic field generation in planets and large moons [9, 41]
- Whole-Mars Decompression Dynamics [30]
- Hydrogen geysers on Mercury [11] and Mars [30]

Even in light of all of these new concepts and discoveries one cannot unambiguously ascertain the Moon's origin. Pieces of the puzzle are still missing which, I suspect, might be found by better understanding details of Earth's origin.

The Moon's maria are similar to flood basalts on Earth [227], two flood basalts in particular, the Deccan Traps [228, 229] and the Siberian Traps [230, 231].

Helium data from the Deccan Traps in India [139] that formed 65 million years ago and from the Siberian Traps [138] that formed 250 million years ago bear the isotopic fingerprint of having been produced by Earth's central nuclear fission georeactor [6]. Helium trapped in contemporaneous basalt floods that form the Hawaiian Islands and Iceland likewise bear the isotopic signature of georeactor-produced helium [137] (Figure 3.7). Moreover, thermal structures or heat channels beneath the Hawaiian islands and

Iceland have been seismically imaged as extending all the way to the top of Earth's fluid core [135, 136].

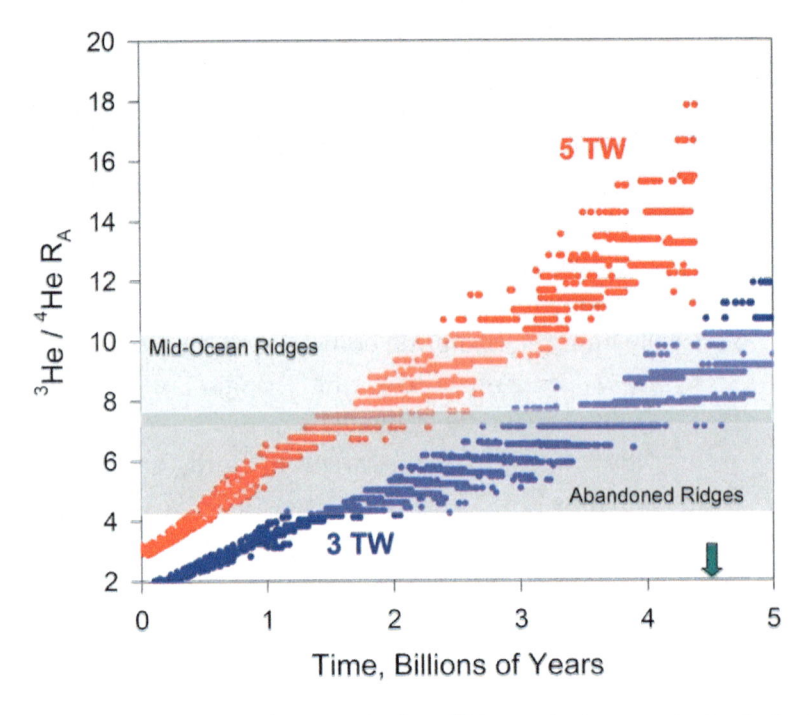

Figure 3.7. Fission product ratio ^{3}He/^{4}He, relative to that of air, R_A, from nuclear georeactor numerical calculations at 5 terawatts, TW, (upper) and 3 TW (lower) power levels [6]. The band for measured values from mid-oceanic ridge basalts is indicated by the solid lines. The age of the Earth is marked by the arrow. Note the distribution of calculated values at 4.5 billion years, the approximate age of the Earth. The increasing values are the consequence of uranium fuel burn-up. Icelandic deep-Earth basalts present values that range as high as 50 times the atmospheric value [98].

Two independent lines of evidence support Earth's georeactor existence:
- Calculated georeactor nuclear fission production of ^{3}He/^{4}He ratios are in precisely the range of ratios observed in oceanic basalts [6].

- Geoneutrino (antineutrino) measurements, at a 95% confidence level, at Kamioka, Japan [108] and Grans Sasso, Italy [109], indicate georeactor nuclear fission output energy of 3.7 and 2.4 terawatts, respectively. These fissionogenic energy values are similar to the 3-6 terawatt range employed in Oak Ridge National Laboratory georeactor simulations [6, 21].

Georeactor formation is a natural consequence of density layering in oxygen-starved (highly-reduced) planetary matter [5, 19, 20]. The two-component, self-regulated [33] nuclear fission georeactor assembly is capable of sustained thermal convection in its charged-particle-rich sub-shell, and is ideally suited for magnetic field generation in planets and large moons [8, 9, 122] (Figure 3.8).

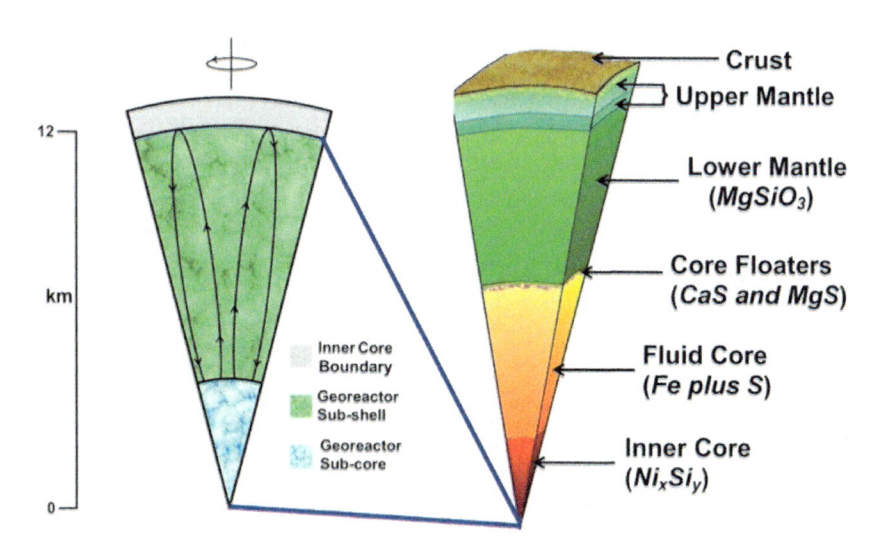

Figure 3.8. Major portions of the Earth's interior from [25] based on [5-8, 13, 19-21, 226] and the relationships shown in Table 1 of Chapter 2. Also, schematic representation of Earth's nuclear fission georeactor with planetary rotation and fluid motions are indicated separately; their resultant motion is not shown. From [30].

Fissionogenic heat produced by the georeactor's nuclear sub-core is transferred via convection in the nuclear waste sub-shell to the inner-core heat sink and then to the larger fluid-core heat sink [13].

A portion of the georeactor produced heat is channeled to Earth's surface hot-spots [10], e.g., Hawaii and Iceland, where its georeactor origin is indicated by the high relative ^{3}He/^{4}He ratios observed [232] and seismically imaged heat channels extending to the top of the core [135, 136].

New Paradigm Origin of Lunar Maria: As noted above, the massive basalt floods, the Deccan Traps and the Siberian Traps were driven by georeactor-produced heat as indicated by the high relative ^{3}He/^{4}He ratios of their occluded helium [138, 139]. These terrestrial basalt floods suggest to me that the lunar maria basalt floods might have similar origins driven by the Moon's nuclear fission "lunar-reactor."

Although the Moon currently has no internally generated magnetic field, remanent magnetization of some of its surface material is indicative of an ancient internally-generated magnetic field [233, 234]. That implication is consistent with the magnetic fields produced by central nuclear fission reactors in many planets and large moons [9, 12].

Earth's formation at high temperatures and high pressures from within a giant gaseous protoplanet resulted in its inner 82%, i.e. the core and lower mantle, being oxygen starved (highly reduced) [12, 32, 122]. Consequently, some of the elements that have a high affinity for oxygen were partially occluded in the iron-alloy core [10, 81]. When thermodynamically feasible, these elements precipitated and separated from the alloy [193]. Uranium, presumably as a sulfide, settled to Earth's center. Silicon combined with nickel settled to form Earth's nickel silicide inner core. Calcium and magnesium combined with sulfur and floated to the top of Earth's core [235, 236].

In 1993, I employed Fermi's nuclear reactor theory [85] to demonstrate the feasibility of a nuclear fission georeactor at Earth's center [5, 19, 20]. In 2001, Dan Hollenbach and I published the results of nuclear georeactor simulation results performed at Oak Ridge National Laboratory [21]. These results showed that the georeactor could operate over Earth's lifetime as a fast neutron breeder reactor. Moreover, subsequent measurements showed that georeactor-produced helium isotope ratios matched helium from terrestrial deep source lavas [6], as pointed out above.

Thus, it is reasonable to suspect that, when the Moon's nuclear fission reactor was in operation, it would provide the heat channeled to the lunar surface to produce the Moon's flood basalts, i.e. the maria.

The location of the lunar-reactor, however, is not at the Moon's center, but at the Moon's center of mass, which is displaced 2 km toward the Earth-facing side [237]. That displacement toward the Earth-facing side in concert with Earth's tidal pull [238], I posit, is principally responsible for driving the maria-basalt floods toward the Earth facing side of the Moon [224].

In principle, it should be possible to verify the correctness of this concept as an explanation for near-side maria bias by measuring the helium isotopes of maria basalt samples taken from depths sufficient to be unaffected by solar wind implanted helium.

When first inspired to look into the problem of the Moon's near-side/far-side maria disparity by Dorion Sagan's question, I was fortunate to have already related, logically and causally, much understanding of Earth's origin, composition, and concomitant geological behavior. This understanding was crucial, because in nature events are rarely independent, but usually are connected to other events and circumstances. Thus, many scientific discoveries can result from a logical progression of understanding, connecting step-by-step events in nature that are related logically and causally, and securely anchored to the properties of matter and radiation.

Despite all of the insights and discoveries made so far, there are more details to understand about Earth's formation which, when made, will presumably lead to an understanding of the Moon's origin.

Mechanism of Solar Activity Triggering Earthquakes, Volcanoes and Geomagnetic Reversals and Excursions: Although long suspected, recently published evidence points to activities on the sun provoking earthquakes [239-246] and volcanic eruptions [247, 248]. However, as noted by Novikov et al. [243], "*The main problem with this research is a lack of physical explanations of a mechanism of earthquake triggering by strong variations of space weather conditions*".

The "*lack of physical explanations*" results from the scientific community adhering to false geophysical paradigms while ignoring or attempting to suppress contradictions thereto [236].

There is widespread belief that our solar system planets formed according to the planetesimal theory [249] despite evidence to the contrary: Earth formed mainly as a consequence of the protoplanetary theory, and only minimally by the planetesimal theory [32].

The currently popular idea of geomagnetic field generation by a convection-driven dynamo mechanism in Earth's fluid core [250] is flawed because thermal convection there is physically-impossible for two reasons [10]: First, due to compression from the weight above, the bottom of the fluid core is 23% denser than the core-top. The small decrease in core-bottom density from thermal expansion (less than 1%) is insufficient to make the core top-heavy as required for convection [251]. Second, for sustained convection, heat brought to the core-top must be quickly removed, a physical impossibility as the core is surrounded by an insulating silicate blanket, the mantle, that has significantly lower thermal conductivity, lower heat capacity, and greater viscosity than the Earth's core.

There are problems with plate tectonics theory: Mantle convection, which is a critical necessity for plate tectonics, is physically-impossible for the following reason: Because of compression by the weight above, the bottom of the mantle is 62% denser than the surface crust [252]. Decreasing mantle-bottom density by thermal expansion (less than 1%) cannot make the mantle top-heavy as required for convection [10, 251]. Additionally, other *ad hoc* assumptions are necessary to make plate tectonics seem to describe geological observations. For example, mountain ranges that predate the assumed collision-formation of Pangea, require the assumption of fictitious supercontinent cycles [41]. Also, inherent errors in geomagnetic paleolatitude determinations [134] lead to false interpretations, for example, rocks from Vancouver Island, Canada thought to have formed in Baja California, Mexico [133].

Trying to pose said *"physical explanations"* on the basis of such a flawed understanding of solid-Earth geoscience is like trying to navigate to a series of addresses in London using an Istanbul city map. However, there is a logical and causally related basis for said *"physical explanations"* that derives from Earth's initial formation as a Jupiter-like gas giant. Codified as *Whole-Earth Decompression Dynamics* [7, 26], my new paradigm replaces plate tectonics.

Whole-Earth Decompression Dynamics, the underlying basis of most geology, geophysics and surface phenomena, is predicated upon the understanding that Earth had fully condensed as a Jupiter-like gas giant when the sun's thermonuclear reactions ignited and the resulting T-Tauri solar winds stripped the ices and gases from Earth's surface [6, 7, 12, 26, 129]. The internal energy sources were a consequence of Earth's protoplanetary formation.

Two powerful energy sources follow from Earth's protoplanetary origin, a central nuclear fission breeder reactor and the potentially much more powerful stored energy of protoplanetary compression. There is an intrinsic relation between the two that is manifest in connection with their response to changes in solar activity and geodynamic consequences.

Earth's Nuclear Fission Georeactor: Earth's condensation from within a giant gaseous protoplanet resulted in its inner 82% existing in a highly-reduced state of oxidation. Because of its oxygen-poor environment, uranium concentrated in the fluid core, instead of in mantle silicates. The uranium precipitated and settled at the center of Earth where it functions as a self-regulating nuclear fission breeder reactor, called the georeactor [5-8, 13, 19-21, 226]. If Earth's magnetic field is generated by a convection-driven dynamo, a magnetic amplifier, as suggested by Elsasser [50], it is produced by the georeactor [8, 13], not in the Earth's fluid core where convection is physically impossible [10].

Georeactor formation is a natural consequence of density layering in oxygen-starved (highly-reduced) planetary matter [5, 19, 20]. The two-component, self-regulated [33] nuclear fission georeactor assembly is capable of sustained thermal convection in its charged-particle-rich sub-shell, and is ideally suited for magnetic field generation in planets and large moons [8, 9, 122].

Fissionogenic heat produced by the georeactor's nuclear sub-core is transferred via convection in the nuclear waste sub-shell to the inner-core heat sink and then to the larger fluid-core heat sink [13]. This process maintains the adverse temperature gradient necessary for thermal convection [251].

The two-component structure of the georeactor provides a natural means of self-regulation. The georeactor sub-shell consists of uranium and radioactive waste, namely, fission fragments and nuclear decay products which are reactor poisons. If, in the microgravity region near Earth's center, the sub-shell components were of uniform density, the reactor poisons would consume a sufficient quantity of neutrons to prevent sustained nuclear fission. Uranium, the densest substance settles out from the less-dense sub-shell and engages in nuclear fission, which disrupts the georeactor assembly. Eventually a steady state is reached wherein the amount of fission energy produced balances the uranium precipitation and the energy transferred to the inner core by convection [33], illustrated in Figure 3.9.

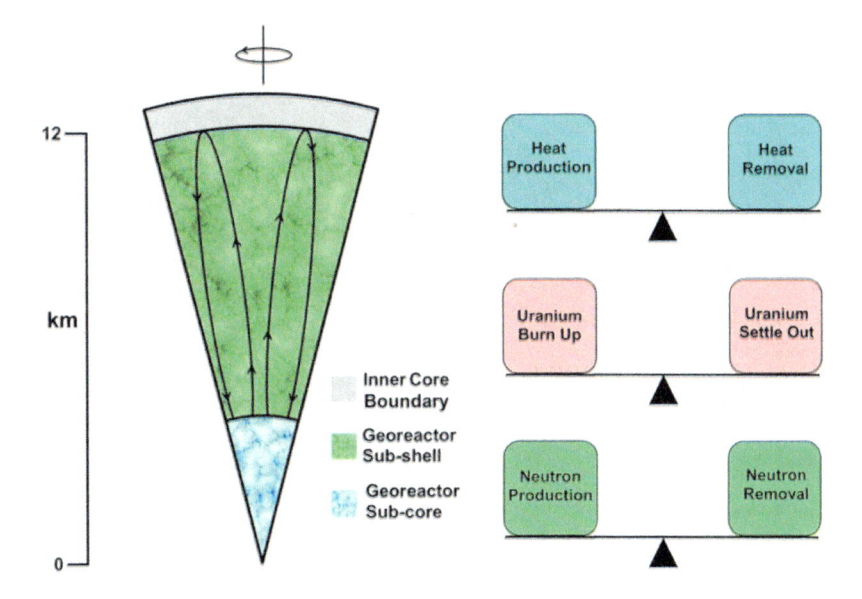

Figure 3.9. Schematic representation of Earth's georeactor, not to scale, with non-resultant planetary and fluid motions indicated separately (left) and (right) representations of the balances that must be maintained for stable georeactor operation. From [33].

The geomagnetic field, I posited, is produced by sustained convection in the radioactive waste sub-shell [5, 6, 8, 13, 18, 20, 122, 235]. The geomagnetic field has been stable, without reversals, for periods longer than 20 million years [253, 254], although more frequent polarity reversals and excursions occur. Clearly, disruptions in georeactor sub-shell convection can lead to geomagnetic field collapse, for example caused by [33]:
- Major trauma at Earth's surface, such as asteroid impact or
- Induced electrical current into georeactor caused by changes in space weather.

Disruption of georeactor sub-shell convection could result by energy from the solar wind transferred via the geomagnetic field into the georeactor by Faraday's law of electromagnetic induction [48]. A simple apparatus, illustrated schematically in Figure 3.10, demonstrates the principle of electromagnetic induction.

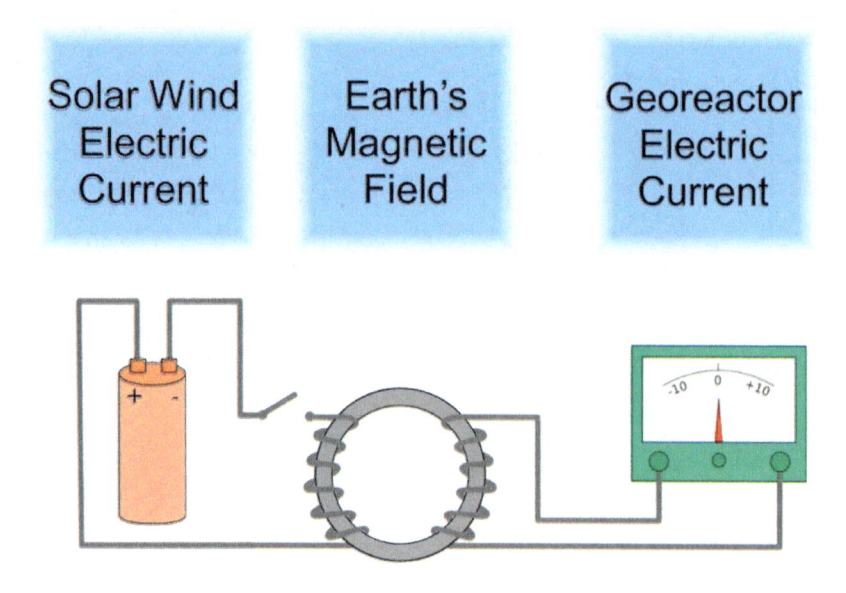

Figure 3.10. Schematic diagram of an apparatus for demonstrating the principle of electromagnetic induction and their corresponding components in nature. From [33].

When the switch in Figure 3.10 is closed, the galvanometer displays only a momentary pulse. When the switch is opened, the galvanometer displays a momentary pulse in the opposite direction. Only a *changing* electrical current can be transferred through electromagnetic induction. The blue boxes in this figure illustrate components in nature that correspond to the schematic electrical components indicated [33].

The solar wind comprises an electrical current of charged particles that stream from the sun. *If the solar wind were constant, no electrical current would be induced into the georeactor.* Exceptionally large changes in the solar wind or in the ring current of charged particles trapped in Earth's magnetosphere or in the cosmic ray flux, however, will cause electrical current to be induced into the georeactor sub-shell producing ohmic heating, diminishing sub-shell convection, and potentially leading to geomagnetic field collapse with concomitant magnetic excursion or reversal [33].

Diminishment of georeactor sub-shell convection may result in a spike of georeactor nuclear fission energy output due to additional uranium settling-out, even if not sufficient to cause a magnetic reversal or excursion [33].

The stored energy of protoplanetary compression is the primary energy source for Earth's decompression. However, for decompression to progress without cooling and impeding decompression, the lost heat of compression must be supplied by georeactor nuclear fission. In addition to doing work against gravity, the stored energy of protoplanetary compression heats the base of the crust by a process known as *mantle decompression thermal tsunami* [123]. Decompression beginning within Earth's mantle propagates outward like a wave through silicates of decreasing density until it reaches the rigid crust where compression and compression-heating takes place. That compression-heating is the heat source for the geothermal gradient as well as for other surface phenomena including shallow-source volcanoes.

The mechanism for changes in solar weather triggering earthquakes and volcanoes is as a multi-stage amplifier. A change in the charged particle flux impinging the Earth's magnetic field induces electric current into the georeactor, which causes ohmic heating, which disrupts sub-shell convection, which results in extra uranium settling-out, which causes a burst of nuclear fission energy, which replaces some of the lost heat of protoplanetary compression, which causes a burst in whole-Earth decompression, which results in a burst of heat emplaced at the base of the crust and/or Earth's surface experiencing a bit of decompression-driven movement, the extent of which is a function of the degree of sub-shell convection disruption.

This mechanism is applicable to solar weather triggering earthquakes and volcanoes as well as posing an explanation for the sometimes observed geomagnetic reversals associated with major geophysical events, such as basalt lava floods [138, 139]. Further investigation should better connect geophysical events to geomagnetic markers.

Whole-Mars Decompression Dynamics: For nearly four decades, NASA-funded individuals have misled the scientific community and the public, effectively replacing scientific ethics with the political behavior of NASA's organizational culture [255]. In the process, those scientists, hand-in-hand with NASA officials, have crippled NASA's ability to understand its own observations and contributed in a major way to the dumbing-down of American science and science education.

An article in the May 2, 1995 issue of *Eos, Transactions, American Geophysical Union*, entitled "Neptune's Nemesis", described observations of a new dark spot in the atmosphere of the planet Neptune. In addition to having a historical error, the article failed to represent to the geophysics community the significance of the observation with respect to possible on-going changes in the planetary driving-energy source. I responded with a brief, 500 word manuscript retort. In submitting the paper to *Eos*, I specifically requested that the manuscript not be sent to NASA's Jet Propulsion Laboratory (JPL) or to the California Institute of Technology (Caltech) that operates JPL for NASA because of a possible institutional conflict of interest.

Most publishers of scientific articles, including at that time the American Geophysical Union (AGU), publisher of *Eos*, have policies clearly stating that editors should avoid real or perceived conflicts of interest. But in blatant contradiction to AGU policy on the avoidance of real or apparent conflict of interest, I was told by the AGU managing editor that an employee of JPL would serve as section editor and that my only other option was to withdraw the article. Not surprisingly, said JPL employee demanded as a condition for publication that I remove all mention of the significance of the observations with respect to possible on-going changes in the planetary driving-energy source which included references to my published work on planetary nuclear fission reactors. There was no legitimate basis for such a demand. This was clearly unwarranted science suppression by an employee of NASA's JPL.

To the best of my knowledge, no NASA-grant recipient has ever cited my publications on planetary nuclear fission reactors [6, 8, 9, 13, 18-20, 29, 33, 86, 122, 255], even though the subject is relevant to virtually all of NASA's planetary investigations. Ignoring potential scientific advances is not a practice of sound scientific reasoning, but misleads the scientific community and blocks further advances and better understandings. Countless billions of dollars of taxpayer provided funds, instead of advancing NASA's science, has been spend promulgating NASA's non-science "storyline." NASA's interpretation of Mars is just one striking example.

In the following I present a fundamentally different Mars paradigm that explains Martian features in a logical way, causally related to the processes operant during the early solar system whose consequences are manifest on other planets.

The prime impetus of scientific investigations of Mars is the search for water, as a potential harbinger of life and as a resource for human exploration.

Presently, only traces of water are present on Mars' surface, although observational evidence suggests the presence of copious amounts of surface water on Mars during its early history. What circumstances initially allowed water's presence, and then subsequently caused its near-absence? These are the questions I address here, however, not by considering Mars as an entity unto itself, but in light of evidence related to the formation of the terrestrial planets generally, and Earth particularly, as set forth in my new indivisible solar system paradigm [12]. That indivisible paradigm accounts for the differences observed among the terrestrial planets and accounts for the asteroid belt as well.

In 1944, Eucken [190] employed thermodynamic considerations to investigate Earth's formation from within a giant gaseous protoplanet. Notably, he discovered that at high pressures Earth's core of molten iron would condense before mantle silicates from an atmosphere of solar composition. Complete condensation would yield a Jupiter-like layer of gases and ices.

In 1976, Suess and I [67] confirmed Eucken's calculations and demonstrated that primordial condensation at high pressures and high temperatures would yield a condensate with a state of oxidation similar to primitive enstatite chondrites, provided the condensate was separated and prevented from further reaction with primordial gases at lower temperatures. Subsequently, I connected Earth's condensation to the circumstances described by Eucken [190] and Suess and I [67] by relating mass-ratios of mineralogically determined parts of a primitive enstatite chondrite to geophysically determined parts-ratios of the Earth [10, 68, 82, 193].

For decades, the abundances of major elements in chondrites have been expressed in the literature as ratios, usually relative to silicon (E_i/Si) and occasionally relative to magnesium (E_i/Mg). By expressing Fe-Mg-Si elemental abundances as atom (molar) ratios relative to iron (E_i/Fe), as shown for comparison in Figure 3.11, I discovered a fundamental relationship bearing on the origin of ordinary chondrite matter [185].

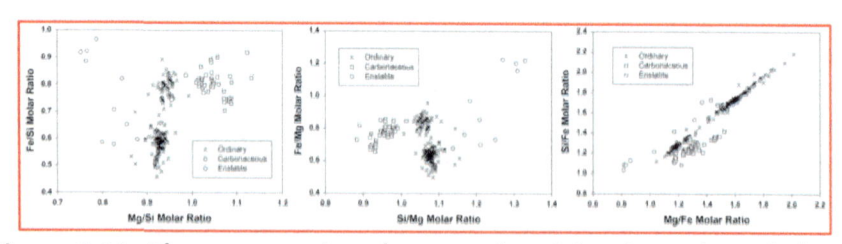

Figure 3.11. **The same major element chondrite data plotted three different ways. The plot on the right, originated by me, shows a relationship that is not evident in the other plots.**

The rightmost plot of Figure 3.11 is presented in greater detail in Figure 3.11 which shows atom (molar) ratios of Mg/Fe vs. Si/Fe from analytical data on 10 enstatite chondrites, 39 carbonaceous chondrites, and 157 ordinary chondrites. The well-defined, linear regression lines are evident only when normalized to Fe, not to Si or Mg. The ordinary chondrite points scatter about a line that intersects the other two lines. Points on the ordinary chondrite line can be represented by mixtures of the two intersecting compositions, point A: *primitive*, and point B: *planetary*. For more detail, see [185]. Near points of intersection, 95% confidence intervals are shown.

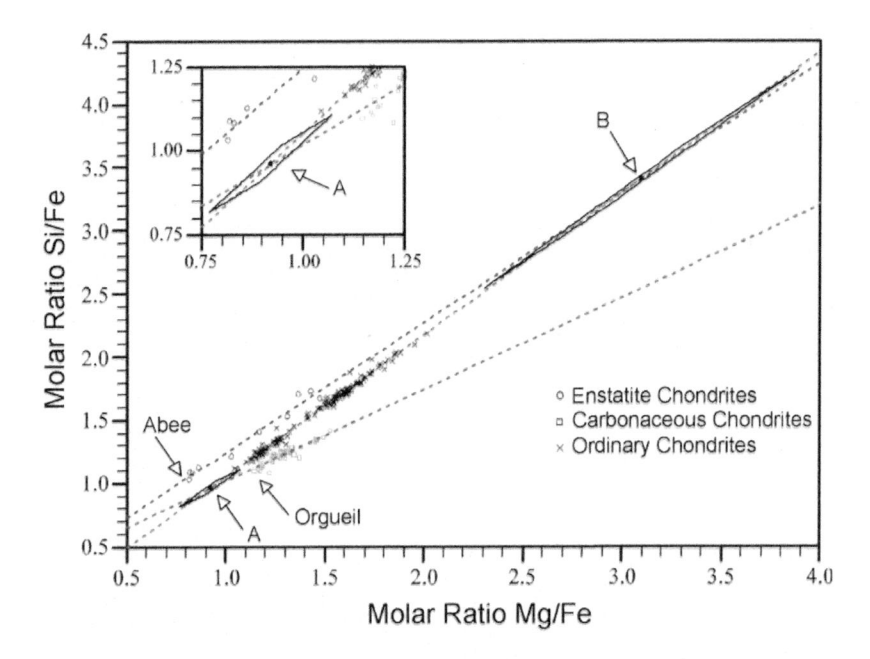

Figure 3.12. Atom (molar) ratios of Mg/Fe vs. Si/Fe from analytical data on 10 enstatite chondrites, 39 carbonaceous chondrites, and 157 ordinary chondrites. Least squares linear regression lines are shown. Near points of intersection, 95% confidence intervals are shown. For references and more detail, see [185].

The relationship I discovered, shown in Figure 3.12, implies that ordinary chondrites were derived from mixtures of two components, representative of two other types of matter, designated *primitive* and *planetary* and defined by the intersecting points along the ordinary chondrite line. The ordinary chondrites consist of mixtures of a relatively undifferentiated carbonaceous-chondrite-like *primitive* component and a partially differentiated enstatite-chondrite-like *planetary* component where it molar (atom) iron content is only one third that of its magnesium and its silicon content.

The *planetary* component, I posited, was the partially differentiated matter stripped from Mercury's protoplanet by the T-Tauri super-intense solar winds where, in the region between Mars and Jupiter, it fused with in-falling primitive matter [185]. The ordinary-chondrite parent matter thus formed populated the asteroid belt (Figure 3.13) and added a veneer that fell onto the outer portion of Earth, and to a greater relative degree, onto Mars.

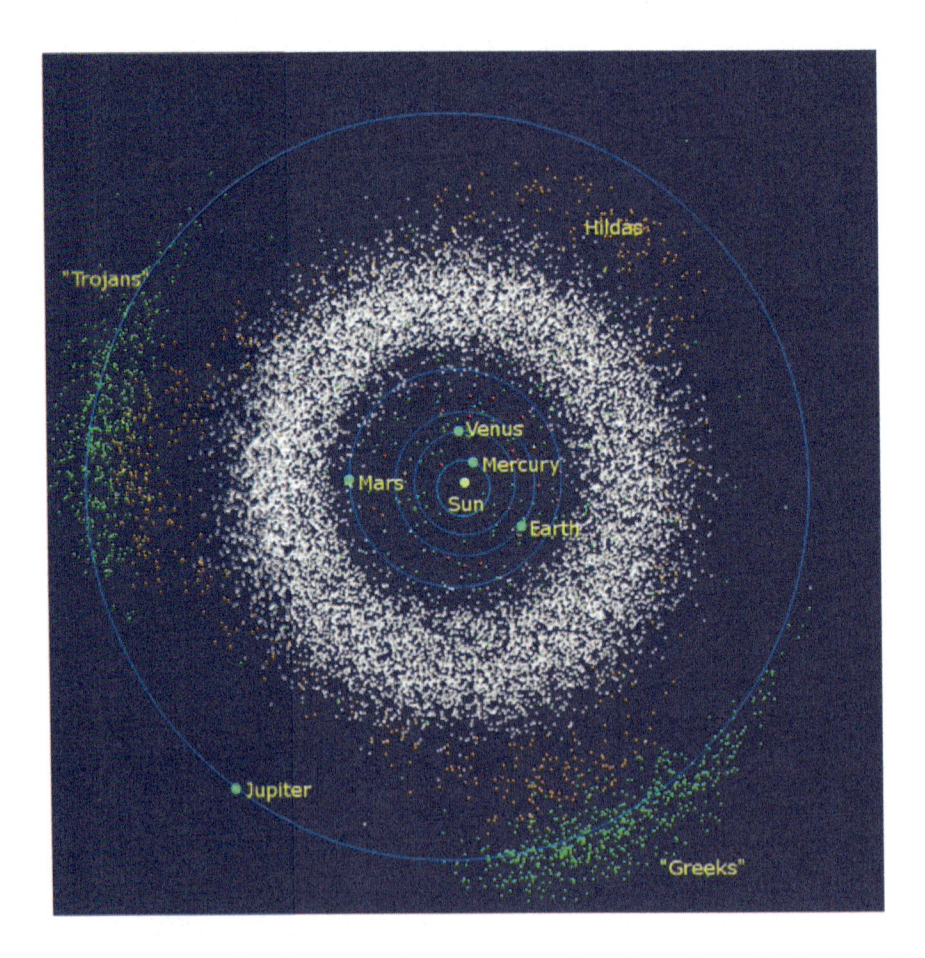

Figure 3.13. The inner solar system showing the plethora of asteroids. From [225].

Several lines of evidence validate the protoplanetary theory of solar system formation [32]. Although the popular planetesimal theory does not account for solar system formation, some of its elements added a veneer of oxidized material to the outer portions of Earth, especially oxidized iron which is critical for the development of life. The high oxidized iron content of the Martian regolith, giving Mars its red color (Figure 3.14), suggests considerably greater "veneer" additions due to Mars' relatively closer proximity to the asteroid belt.

Figure 3.14. True color image of Mars taken by the OSIRIS instrument on the ESA *Rosetta* spacecraft during its February 2007 flyby of the planet. Courtesy of ESA and MPS. CC BY-SA 3.0 IGO.

New Paradigm: Whole–Mars Decompression Dynamics: Evidence that Mars had an internally generated magnetic field early in its lifetime [256-258] is evidence that interior portions of Mars experienced protoplanetary condensation. In the highly reducing solar matter environment at high pressures and high temperatures, molten iron and all of the elements dissolved in it, including uranium, is the first major condensate to rain-out directly forming the core. Planetocentric nuclear fission reactors, self-regulating and producing planetary magnetic fields, are the inevitable consequence of protoplanetary condensation, including Mars.

Evidence of Whole-Mars Decompression Dynamics is not generally as conspicuous as the mountain ranges characterized by folding on Earth or the circum-Pacific trenches. However, one extremely prominent feature of Mars represents an outstanding example. Valles Marineris, in the southern hemisphere, is a set of partially in-filled secondary decompression cracks 150-2200 km long, 75-150 km wide, and at least 5-10 km deep [259-261]. As a consequence of whole-Mars decompression increased diameter, these decompression cracks necessarily formed to increase the surface area correspondingly.

Figure 3.15 is a mosaic of the Valles Marineris hemisphere of Mars projected into point perspective, a view similar to that which one would see from a spacecraft, according to NASA. The distance is 2500 km from the surface of the planet, with a scale being 0.6km/pixel. The mosaic is composed of 102 Viking Orbiter images of Mars. The center of the scene (latitude -8, longitude 78) shows the entire Valles Marineris secondary decompression crack system.

Figure 3.15. Mosaic of the Valles Marineris hemisphere of Mars. NASA image.

Figure 3.16 is a Mercator map of Mars onto which I added names of prominent, related features.

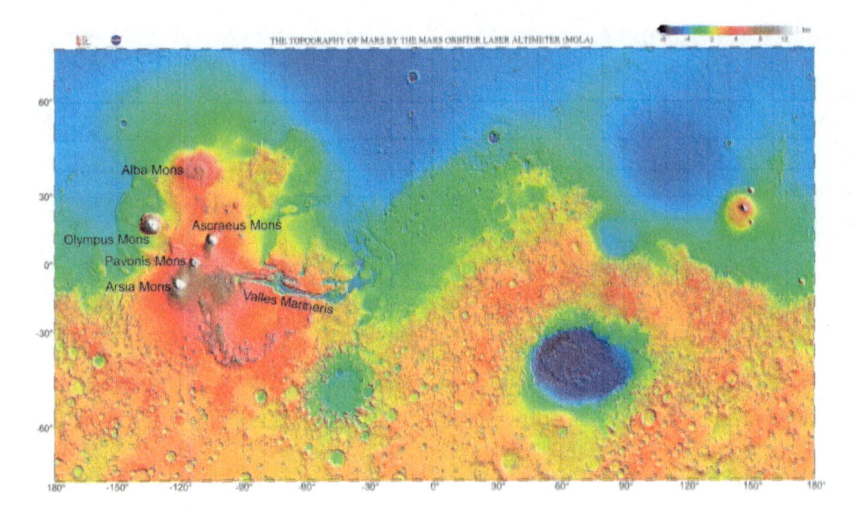

Figure 3.16. Mercator map showing the concentration of volcanoes near Valles Marineris.

Figure 3.17 is a mosaic image of Valles Marineris – colored to resemble the Martian surface – derived from the Thermal Imaging System (THEMIS), a visible-light and infrared-sensing camera on NASA's Mars Odyssey orbiter. Produced from more than 500 daytime infrared photos, the mosaic shows the whole valley in more detail than any previous composite photo. The smallest details visible in the image are about the size of a football field, 100 meters.

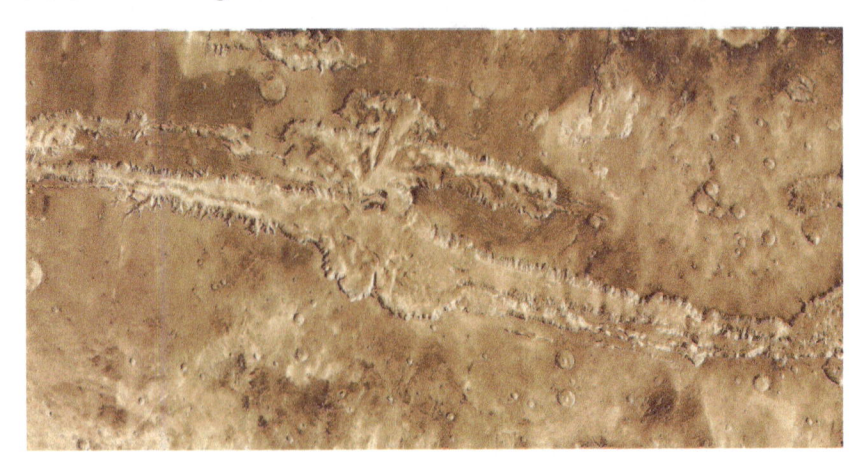

Figure 3.17. Mosaic image of Valles Marineris, artificially colored, constructed from more than 500 daytime infrared photos. NASA image.

Just as on Earth, during whole-Mars decompression, (1) surface cracks form that are subsequently in-filled to increase surface area, and (2) surface curvature adjustments must be made.

The upper image in Figure 3.18 is a magnified view of a section of Figure 3.17 showing in greater detail the perimeter-features common to the partially in-filled secondary decompression cracks of Valles Marineris. The inset in the upper image from Figure 2.16c serves as a reminder of two means of surface curvature adjustment. To date, no evidence of Martian mountains characterized by folding has been reported. However, the common perimeter-features of Valles Marineris, I posit, are perimeter-tears like those of the Norwegian fjords [14] as shown by the lower images of Figure 3.18. Those perimeter-tears are indications of Martian surface curvature adjustments.

Figure 3.18. Upper: Magnified section of Figure 3.17, Valles Marineris, showing circum-perimeter tears, with inset from Figure 2.16c illustrating mechanisms for surface curvature adjustments; Lower: USGS/NASA satellite view of the northern portion of Norway showing fjords and map of Norway showing fjords, from [14].

Figure 3.19 shows two rimless pits located to the northwest of Ascraeus Mons. The pits are 180 and 310 meters in diameter. The associated wispy, dark material appears to have blown out of the pits. Although the Martian pits are considerably larger and far fewer than the pits on Mercury discovered by the Project Messenger mission [194], I suspect that they may be of similar origin, namely the result of hydrogen geysers [11]. Molten iron, which dissolves copious amounts of hydrogen, is the first major condensate during protoplanetary formation. Eventually, when the planetary core solidifies, the hydrogen is exsolved and rushes to the surface. Along the way the hydrogen reduces iron sulfide to iron metal, which is blown out and deposited at the surface. On Mercury, which is devoid of atmospheric winds, the iron is deposited around the pits and remains in its reduced state (Figure 3.20). On Mars, presumably the iron is blown downwind and becomes oxidized. In each case, the validity of this concept can be tested by determining whether the deposited material is iron metal and iron oxide, respectively.

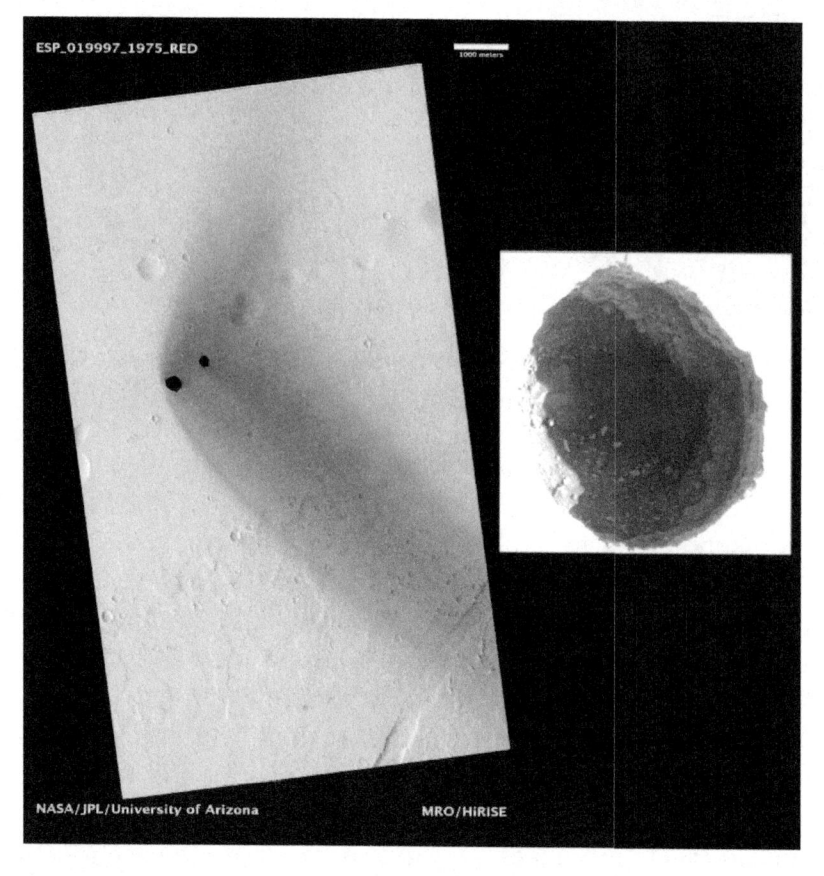

Figure 3.19. Dark rimless pits NW of Ascraeus Mons. Inset: Close up of rimless pit.

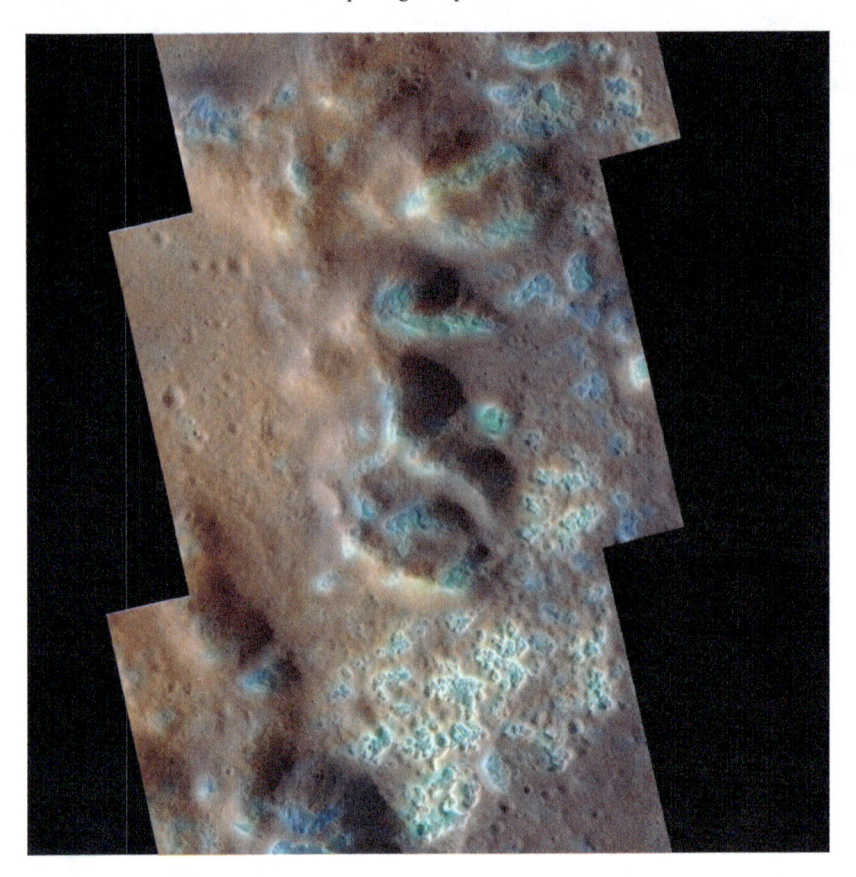

Figure 3.20. Pits surrounded by shiny material on planet Mercury. Repeat of Figure 3.2. From [194].

Where to Look for Martian Organic Compounds: If evidence of extra-terrestrial life is to be found in the solar system, Mars is the most likely place to look. Indeed, the primary focus of space agencies' exploration has been to seek likely regions, such as those revealing evidence of past flowing water [262]. An alternative approach is to seek the sources of Martian methane [263]. In an article published in the *Journal of Petroleum Exploration and Production Technology*, I described the scientific basis and evidence for a "New Concept on the Origin of Petroleum and Natural Gas" that follows from Whole-Earth Decompression Dynamics [147]. Based on these inferences, I suggest appropriate places to look for trapped organic compounds that might have arisen from whole-Mars decompression.

Whole-Earth decompression currently is operating, although less violently than in the past. A notable example is the East African Rift System shown with its associated volcanoes in Figure 3.21. The volcanoes associated with the East African Rift System are powered by heat channeled from the georeactor, as known by the isotopic signature helium that makes its way to the surface rocks through those heat channels. It is worthwhile to enquire whether the heat that produced the volcanoes associated with Valles Marineris is likewise of nuclear reactor origin, as can be ascertained by making helium isotopic measurements on samples shielded from solar wind implanted helium. If so, then it might be prudent to seek organic deposits where life might have originated that are associated with the Valles Marineris region. Why? The East African Rift System (Figure 3.21) and the Siberian Traps (Figure 3.22) have extensive oil and natural gas deposits [147].

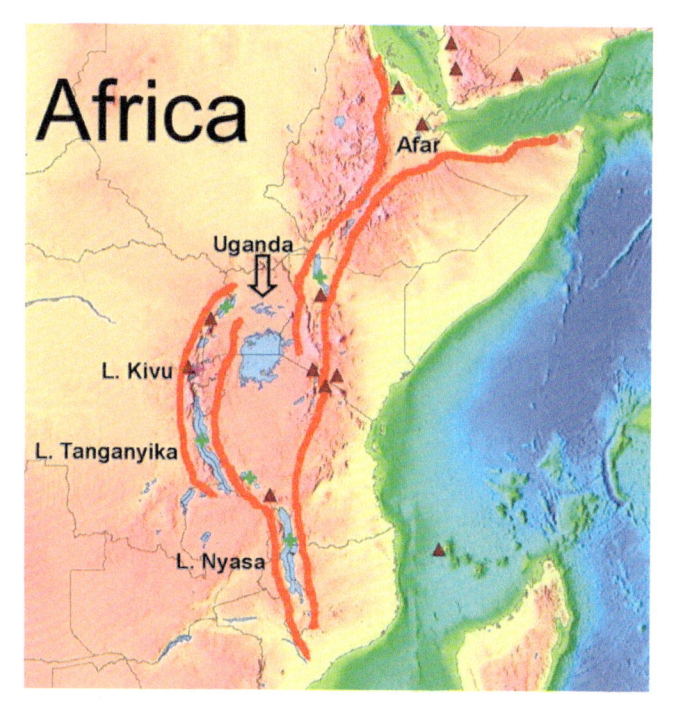

Figure 3.21. Northeastern portion of Africa. Red lines show the major decompression cracks comprising the East African Rift System. Active volcanoes are indicated by maroon triangles, oil discoveries by pluses [147, 264].

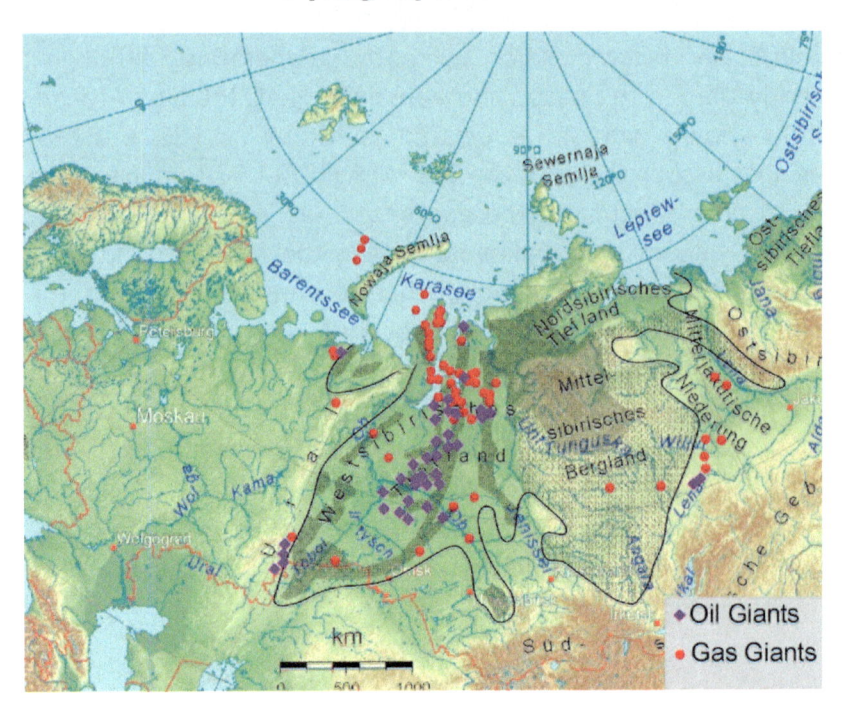

Figure 3.22. Map, courtesy of Jo Weber, showing the extent of the Siberian Traps as based upon estimates derived from Masaitis [265]. Circles show major gas fields; diamonds show major oil fields; data from Horn [266]. From [147].

Mars Mantle Decompression Thermal Tsunami: Whole-Earth decompression progressed far more energetically than whole-Mars decompression. Based upon relative planetary masses, the Martian protoplanetary kernel was much smaller than that of Earth, less than 11% as massive. The early demise of Mars' magnetic field [256-258] is consistent with a much smaller planetocentric nuclear fission reactor. The much less extensive Whole-Mars Decompression Dynamics is also a consequence of its being compressed by a less massive shell of ices and gases, less than about 11% the mass of Earth's primordial gaseous shell.

Since 1939, scientists have been measuring the heat flowing out of Earth's continental crust [267, 268] and since 1952, the heat flowing out of ocean floor basalt far removed from mid-ocean ridges [269].

When the first heat flow measurements were reported on continental-rock, the heat was assumed to arise from radioactive decay. But then later, ocean floor heat flow measurements were made and what a surprise! There is more heat flowing out of the ocean floor than out of continental-rock. This was an enigma; ocean-floor basalt contains a much smaller percentage of radioactive elements than continental-rock. The solution to that enigma is the Whole-Earth Decompression Dynamics process called mantle decompression thermal tsunami.

Albeit controversial, there is evidence that during an early period on Mars there existed copious quantities of liquid water that was subsequently lost [270, 271]. However, so far no one has published a holistic logical explanation for this postulation. How can the planet be initially warm and wet then change to cold and dry? The answer, I posit, is due to the Whole-Mars Decompression Dynamics process of mantle decompression thermal tsunami that while operating (1) heated the Martian surface and (2) provided a thermal barrier that prevented water from percolating too deeply underground. When mantle decompression thermal tsunami ceased, heat flowing from within the planet to the Martian surface subsided, and water encountered no thermal barrier to downward percolation. Presumably, this event was more or less contemporaneous with the demise of the planet's magnetic field which began the solar wind caused erosion of the Martian atmosphere [272].

Recapitulation: Whole-Mars Decompression Dynamics, like Whole-Earth Decompression Dynamics, is based upon evidence that planets condensed from giant gaseous protoplanets which established their highly reduced internal compositions, their planetocentric nuclear fission reactors, and ultimately resulted in their formation as gas giants, surrounded by shells of ices and gases amounting to about 300 times the rocky planet mass. The resulting compression stored within the rocky planets is a powerful energy source, the energy of protoplanetary compression. The super-intense T-Tauri phase solar winds accompanying thermonuclear ignition of the sun stripped away the gaseous envelopes thus beginning the process of whole-planet decompression.

Whole-Earth decompression progressed far more energetically than whole-Mars decompression. Based upon relative planetary masses, the Martian protoplanetary kernel was much smaller than that of Earth. The early demise of Mars' magnetic field is consistent with a much smaller planetocentric nuclear fission reactor. The much less extensive Whole-Mars Decompression Dynamics is also a consequence of Mars being compressed by a less massive shell of ices and gases. Nevertheless, surface evidence, notably Valles Marineris, is consistent with whole-Mars decompression.

Perhaps the most important consequence of Whole-Mars Decompression Dynamics is, through the process of mantle decompression thermal tsunami, the stored energy of protoplanetary compression also heated the crust and provided a thermal barrier to the downward percolation of water. Presumably contemporaneous with demise of the magnetic field, the whole-Mars decompression process subsided, concomitantly transforming Mars from warm and wet to cold and dry. Whole-Mars Decompression Dynamics provides different possible interpretations of Martian features and events, including the perimeter-features of Valles Marineris being circum-perimeter tears like Norwegian fjords, surface pits being the consequence of hydrogen geysers like on Mercury, and implications where one might hope to find Martian hydrocarbon deposits.

4 ASTROPHYSICAL SCIENCE MISTAKES AND DECEIT

Throughout human history, scientific knowledge sometimes has been a source of enlightenment and other times, an excuse for persecution. Science is all about truth, truth securely anchored to the properties of matter and radiation. The science of our world today, however, has departed from standards of truth and objectivity, and has all too often become an arena for deceit by the science controllers. Yet truth is a fundamentally important human determinant, inextricably connected to the freedom we seek and deem precious.

The importance of the sun for human existence was recognized long ago in ancient cultures, and figured prominently in their religions and cosmologies [273-275]. Yet critical knowledge of our star, the sun, and the implications derived therefrom, which I published in the scientific literature [20, 36, 276], has neither been shared nor disclosed by today's Deep State scientists. But I share knowledge pertaining to the ignition of the sun which is crucial for progress along the path to better understand our Universe.

The Starlight Problem: At the beginning of the 20th century, understanding the nature of the energy source that powers the sun and other stars was one of the most important unsolved problems in the physical sciences. Initially, it was thought that during formation, when dust and gas coalesce and collapse by gravitational attraction, great amounts of heat would be produced. But calculations showed that the energy release would be insufficient to power the sun for as long as life has existed on Earth. Following the discovery of radioactivity by Becquerel in 1896 [277], numerous experiments began to reveal the nature of radioactivity, the atomic nucleus, and nuclear reactions [278].

In 1934, Oliphant, Harteck, and Rutherford [279] discovered thermonuclear fusion reactions, an example of which is illustrated in Figure 4.1.

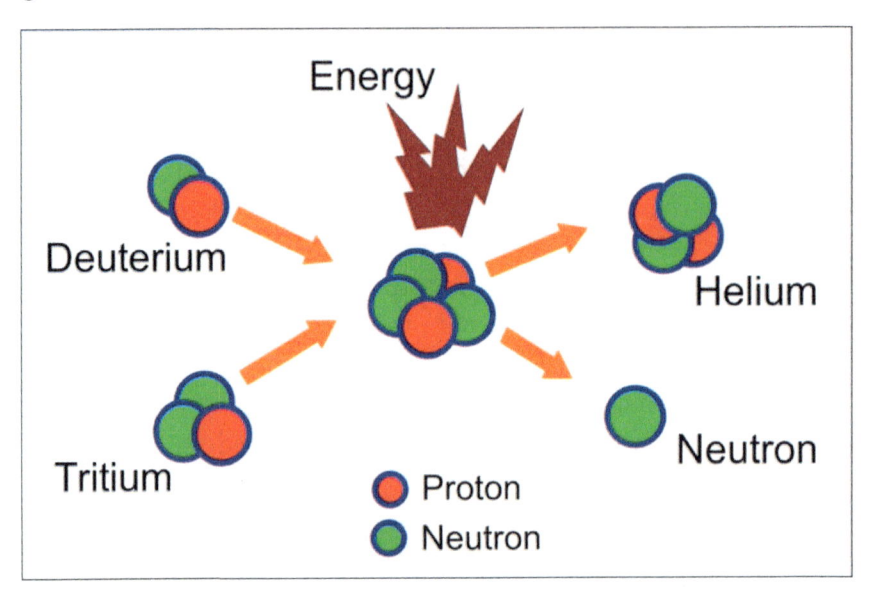

Figure 4.1. Schematic representation of a thermonuclear fusion reaction. The nuclei of light elements, deuterium and tritium, "fuse" to produce helium, a neutron, and a great amount of energy.

Thermonuclear fusion reactions are called thermonuclear because temperatures on the order of 1,000,000°C are required for the nuclei to achieve the very high velocities needed to overcome the electric charge repulsion and get close enough for the nuclei to react. When the fusion reaction takes place, a great quantity of energy is released.

Thermonuclear fusion reactions seemed to be the unknown source of energy that powers the sun and other stars, which contain copious amounts of hydrogen and helium. The scientific development of solar thermonuclear reactions was undertaken by nuclear physicists such as Edward Teller [280] and Hans Bethe [281], whose names would later be associated with the development of nuclear weapons.

By 1938, theoretical investigations on the thermonuclear reactions thought to power the sun and other stars had sufficiently progressed that there seemed to be no longer any question as to the sun's energy source. But as often happens in science, *"the devil is in the details."* In 1938, there was no energy source known that could produce the million degree temperatures necessary to ignite thermonuclear fusion reactions. So, it was just assumed that such temperatures would be produced during star formation when dust and gas coalesce and collapse by gravitational attraction.

Scientists tend to be forward-looking, and rarely look questioningly at circumstances in the past that set them on their present path. That was certainly the case for igniting thermonuclear reactions in stars by gravitational collapse. In 1965, Hayashi and Nakano [282] first showed that gravitational collapse of dust and gas during star formation would not yield the requisite million degree temperatures for igniting thermonuclear fusion reactions. The reason is obvious. Heating a forming star by gravitational collapse of dust and gas is offset by heat radiated from its surface, which is a function of the fourth power of temperature. In other words, TxTxTxT represents a huge loss factor when T=1,000,000°C. But instead of asking "what is wrong with this picture", astrophysicists just made *ad hoc* assumptions, such as a shock-wave induced flare up, or they tweaked model-parameters in attempts to attain the requisite temperatures [283, 284].

The sun is like a hydrogen bomb held together by gravity (Figure 4.2). Both are powered by thermonuclear fusion reactions, and both require temperatures on the order of a million degrees Celcius for ignition.

Figure 4.2. The sun (left) is like an on-going hydrogen bomb (right) held together by gravity.

Both Teller and Bethe made crucial contributions to hydrogen bomb technology. But another critical discovery was made between the time of their work on thermonuclear reactions in the sun and on hydrogen bombs. That discovery, made in December 1938 and published in *Die Naturwissenschaften* in January 1939, was nuclear fission, the splitting of the uranium nucleus [285].

As experimental investigations early in the century revealed, nuclear reactions can be artificially induced by bombarding a target nucleus with neutrons. This may cause the target nucleus to become an entirely different element, changing its element number (proton number) by no more than two. In 1938, however, Hahn and Strassmann [285] bombarded uranium with neutrons, and chemically detected barium, an element about half the proton number of uranium. Hahn and Strassmann had split the uranium nucleus into two parts.

Splitting the uranium nucleus releases an enormous amount of energy and liberates neutrons. These newly released neutrons could split other uranium nuclei, which could split others, and so forth in a chain reaction that is the basis for the atomic (fission) bomb [286, 287] and nuclear reactors [288] (Figure 4.3).

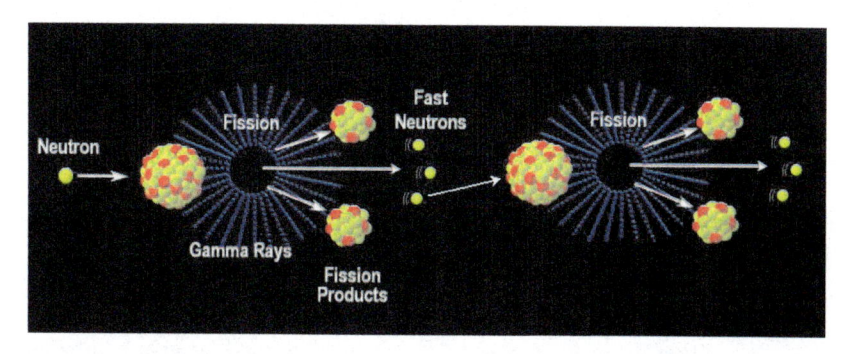

Figure 4.3. Schematic representation of the uranium nuclear fission chain reaction.

Nuclear fission, discovered as war clouds were gathering over Europe in December 1938, immediately became of paramount interest as a potential new weapon of war. That potentiality became a reality with the detonation of atomic (fission) bombs over Hiroshima and Nagasaki in 1945 [287]. Just seven years later, the United States detonated the first thermonuclear fusion bomb, also called hydrogen bomb, on Eniwetok Atoll in the Pacific Ocean [289]. That hydrogen bomb and all subsequent hydrogen bombs have utilized a nuclear fission chain reaction device to ignite their thermonuclear fusion reactions.

Natural Nuclear Fission Reactors: Enrico Fermi formulated nuclear reactor theory [85] and in 1942 constructed the first man-made nuclear fission reactor at the University of Chicago. Producing a nuclear fission chain reaction from naturally occurring uranium required a clever reactor design because readily fissionable U-235 presently comprises only 0.7% of uranium.

In 1956, Paul Kazuo Kuroda applied Fermi's nuclear reactor theory and demonstrated that nuclear fission chain reactions could have occurred in seams of uranium ore two billion years ago when the relative proportion of U-235 was greater [290, 291]. Kuroda later told me that the idea was so unpopular that the only way he managed to get it published was because at the time the *Journal of Chemical Physics* would publish short papers without review. Even in 1956 peer-reviews were being used as a means to suppress publication of scientific advances!

In 1972, French scientists discovered, in a uranium mine at Oklo in the Republic of Gabon in Western Africa [292, 293], the intact remains of a natural nuclear reactor that had operated as predicted by Kuroda [290, 291] (Figure 4.4).

Figure 4.4. Seam of uranium ore in an Oklo natural nuclear reactor zone. Photo courtesy of Francoise Gauthier-Lafaye.

As astronomers first discovered in the late 1960s, three of the giant gaseous planets, Jupiter, Saturn, and Neptune radiate into space approximately twice the energy they receive from the sun and display prominent turbulence [294, 295] (Figure 4.5). The explanation proffered by NASA-funded scientists was that the energy is gravitational [296]. It did not make sense to me that Jupiter should still be collapsing after 4.5 billion years. Reflecting on the problem, I realized that Jupiter has all the ingredients for a planetocentric nuclear fission reactor. I applied Fermi's nuclear reactor theory to demonstrate the feasibility that the internal energy production and atmospheric turbulence in the giant planets is produced by planetocentric nuclear fission reactors. My scientific paper on the subject was published by *Naturwissenschaften* in 1992 [18].

Figure 4.5. Turbulence in the atmospheres of Jupiter, Saturn, and Neptune, but not conspicuous in the atmosphere of Uranus, pictured in the lower left.

Initially, I thought that hydrogen would be necessary to slow neutrons for the nuclear fission chain reaction (Figure 4.3), but quickly realized that hydrogen was not at all necessary. That opened the possibility of central nuclear fission reactors inside other planets and large moons, such as Io, that could power and produce their magnetic fields [5, 6, 8, 9, 13, 18-21, 86]. To my knowledge, over a period of 32 years, NASA-funded scientists have never cited my work on nuclear fission reactors inside planets and large moons, despite its being published in some of the world's foremost scientific journals [5, 6, 8, 9, 13, 18-21, 86, 297].

Thermonuclear Ignition of Stars: Shortly after publishing my demonstration of the feasibility of nuclear fission reactors for the giant planets [18], I started thinking about Jupiter being similar to, but much too small to have become a star. A star is like a hydrogen bomb held together by gravity, and all hydrogen bombs are ignited by their own small atomic (nuclear fission) devices. Is it possible that the thermonuclear reactions in stars are ignited by nuclear fission chain reactions? Could it be that the astrophysics community missed that? That seemed unlikely, especially as both Teller and Bethe had done pioneering work on thermonuclear reactions that power the sun, and both had worked on the development of hydrogen bombs. In fact, Edward Teller is known as the *father of the hydrogen bomb*.

Nevertheless, I headed for the science library and meticulously researched the literature. I found no mention of stellar thermonuclear ignition by nuclear fission in the scientific journals. And to be sure, I even hired a research librarian to search all available online data bases. That computer search still revealed no mention of stellar thermonuclear ignition by nuclear fission. Amazing! Teller and Bethe had neglected to look over their shoulders, neglected to reconsider their previous work in light of the lessons learned from their later work.

I promptly wrote a short scientific article about thermonuclear ignition of stars by nuclear fission chain reactions, which was rejected by several journals before being accepted for publication in the *Proceedings of the Royal Society of London* [20]. One of the rejections was based upon an anonymous reviewer's remark that "Herndon is throwing away forty years of astrophysics." So, what is wrong with that? Science progresses by finding what is wrong with current ideas and correcting them. Scientists should have welcomed my paradigm shift as it affords new insights and opens new possibilities for scientific discoveries [42].

New Insight on the Nature of Dark Matter: It has always been my experience that new insights and discoveries inevitably lead to further new insights and to further new discoveries. In the old, flawed paradigm, stars (except tiny brown dwarfs) are always thought to ignite by gravitational collapse during formation. In my new paradigm, however, stellar ignition requires the presence of very heavy elements, such as uranium or plutonium, to undergo nuclear fission chain reactions. Without fissionable heavy elements, after cooling from contraction, the stars would be dark stars.

Without heat generated by thermonuclear reactions to expand their gas, a dark star the mass of the sun would have a diameter similar to that of Earth. One of the consequences of my new insight on star ignition is that it sheds light on the nature of dark matter [20].

A spiral galaxy, such as shown in Figure 4.6, represents a dynamically unstable assemblage of stars that would hypothetically wrap around its center of rotation, unless it is surrounded by a massive halo of unseen (dark) matter 10-100 times as massive as the luminous stars [298]. In Figure 4.6, this yet unobserved halo of dark matter is illustrated in green. I suggested that the dark matter surrounding luminous galaxies is composed of dark stars, the consequence of stellar non-ignition that results from the absence of fissionable elements. I even pointed out corroborating evidence, namely, the association of low-metal stars in the regions believed populated by dark matter [20].

Figure 4.6. Typical spiral galaxy. The hypothetical green halo shows the region where dark matter is thought to reside, imparting dynamic stability to the luminous configuration of stars [298].

The question of what constitutes dark matter is a subject of active debate in the astrophysics community with a wide range of exotic possibilities discussed, such as hypothetical axions and putative primordial black holes [299]. To the best of my knowledge, in the intervening 30 years since publication of my concept of dark matter consisting of dark stars [20], no astrophysicist has cited my suggestion that zero-metallicity dark stars may account for at least a significant portion of the dark matter in the Universe.

Thermonuclear Ignition of Dark Galaxies: Figure 4.7 is a Hubble Space Telescope deep-field view showing approximately 15,000 galaxies. Two features stand out and beg for explanation. First, among this vast number of galaxies, there are only a few prominent morphologies, suggesting a commonality of formative conditions. Second, a vast proportion of the observable luminous galaxies are flat, not spherical.

Figure 4.7. Hubble Space Telescope deep field photograph showing approximately 15,000 galaxies.

Astronomers have produced a wealth of observations about the matter in the Universe. Astrophysicists attempt to explain these observations, but they are crippled by scientific failings, for example, by making models based upon assumptions instead of making discoveries, by ignoring contradictory concepts, and by accepting without question obtuse ideas. I include among those obtuse ideas the concept that the Universe sprang into existence from a point of nothingness some 13.8 billion years ago, and that at the center of galaxies matter disappears forever into black holes.

From long experience I have learned that nature can usually be understood as operating in logical, causally related ways that do not require unscientific suppositions.

Galaxies are massive assemblages of matter, some containing as many as a billion luminous stars. As matter at the galactic center becomes extremely massive, it does not disappear forever into black-hole nothingness, but instead matter is jetted from the galactic center into space as mono-polar or bi-polar galactic jets (Figure 4.8).

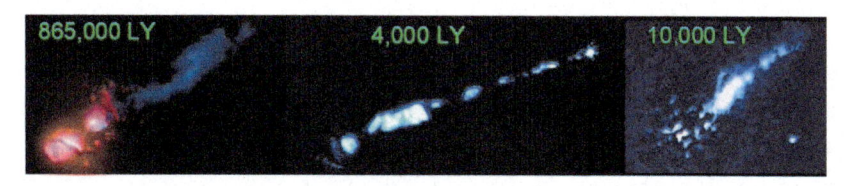

Figure 4.8. Hubble Space Telescope images of galactic jets with their lengths indicated in light years.

As I have published [8, 20, 36, 264, 276], the morphological features and galactic luminous star distributions can be understood in logical, causally related ways. Consider a dark galaxy consisting solely of zero-metallicity dark stars, stars consisting only of hydrogen and helium. As the dark matter coalesces and becomes extremely dense at the galactic center, at some point the galactic center shoots out its first galactic jet. The galactic jet, I contend, seeds any of the dark stars it contacts with fissionable elements, capable of producing nuclear fission chain reactions, thus providing the million-degree temperatures necessary to ignite their stellar thermonuclear fusion reactions [36, 276].

What would a spherical dark galaxy look like after its first galactic jet? Figure 4.9a,b show two examples.

Figure 4.9a, NGC 4676, referred to as the *Mice Galaxies*, are two spiral galaxies. Note the "tail" of the galaxy on the right. This is a line of luminous stars that were ignited when that galaxy sent from its center its first galactic jet, which seeded the dark stars it encountered with fissionable elements that produced nuclear fission chain reactions that provided the million degree temperatures to ignite their thermonuclear fusion reactions, thus turning the dark stars into luminous stars.

Figure 4.9b, UGC 10214, referred to as the *Tadpole Galaxy*, is a barred spiral galaxy in the early stage of luminosity when galactic jets sent from its center first begin to seed its dark stars with fissionable elements which, by nuclear fission chain reactions, ignite the dark stars encountered by the galactic jets.

Figures 5.9c,d show more evolved galactic luminous star distributions that nevertheless display the path of former galactic jets that provided the heavy-element component permitting stellar thermonuclear ignition. And what of the dark matter necessary for dynamical stability of the luminous structures? It is there in the un-ignited portion of the spherical dark galaxies.

Figure 4.9. Hubble Space Telescope image of (a) NGC 4676, *Mice Galaxies*, (b) UGC 10214, *Tadpole Galaxy*, (c) spiral galaxy, M101, and (d) barred spiral galaxy, NGC 1300.

Blacklisting: In 2006, I submitted a short manuscript on the thermonuclear ignition of dark galaxies to *Astrophysical Journal Letters*. I signed the required copyright transfer form, and the manuscript went out for secret "peer-review," but it was rejected without any substantive scientific criticism. So I submitted two other brief manuscripts. The fact that I was never asked to sign the copyright transfer forms for those other two papers prior to review as required, a serious breach of journal policy, was clear indication that my manuscripts were not going to be accorded the fair and impartial consideration that is supposed to be the usual policy of the American Astronomical Society, the journal's sponsor. I complained to the officers of the American Astronomical Society, who never responded, even though the bylaws of the American Astronomical Society (*AAS*) at the time clearly stated: *"As a professional society, the AAS must provide an environment that encourages the free expression and exchange of scientific ideas"*.

Not long after the *Astrophysical Journal Letters* incident, on July 27, 2007, when attempting to post a preprint in the astrophysics category, I found that I had been blacklisted by *arXiv.org*, an author self-posting archive that functions, in theory at least, to preserve and make available to other scientists valuable potential insights that would otherwise be quashed by unethical peer-review and other bureaucratic restrictions and machinations (Figure 4.10). Blacklisting by arXiv.org is a well-known monopolistic mechanism with no recourse that censors legitimate scientific expression and deceives the scientific community and government science-funding agencies [114].

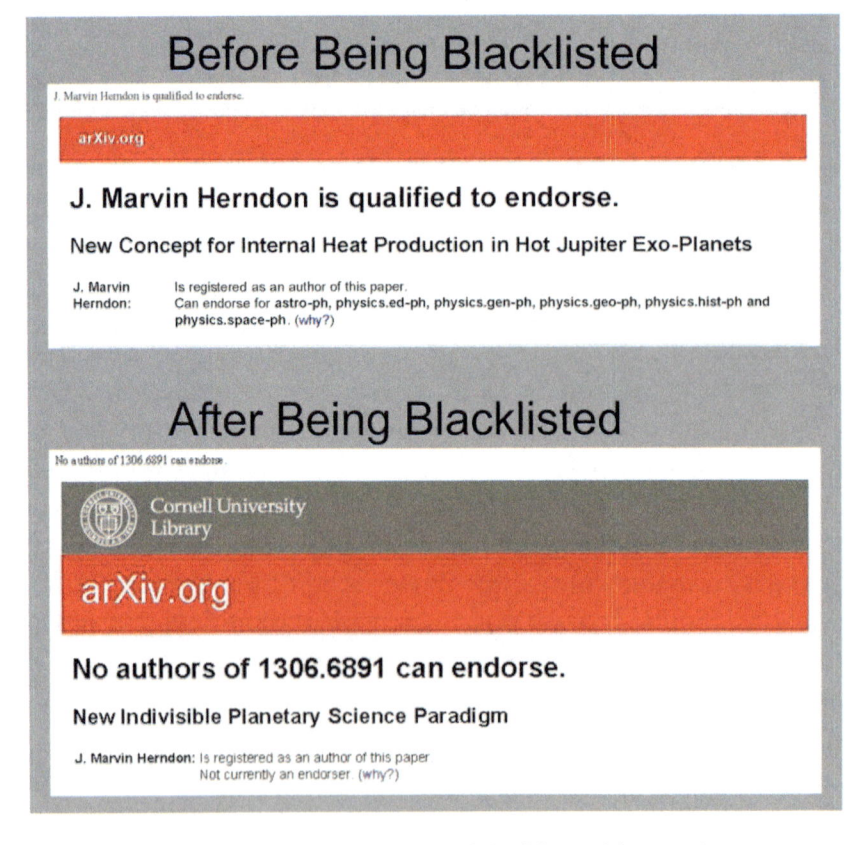

Figure 4.10. Before and after being blacklisted by arXiv.org.

Origin of the Elements: In a 1957 scientific article, entitled "Synthesis of the Elements in Stars," Burbidge, Burbidge, Fowler, and Hoyle [300] proposed that chemical elements are synthesized in stars by a number of processes. Heavy elements, however, were assumed to be solely produced by "rapid neutron capture" during supernova explosions. These ideas are still widely believed [183]. Subsequent observations [20], I posit, lead to a fundamentally different understanding of the origin of the elements [264], which I describe briefly here.

Astrophysicists group stars into two categories based upon their metal content. The association of low-metal stars in the region believed populated by zero-metal stars, i.e., dark stars [20], suggests to me that there exist two *primary* sources of chemical elements. One of the two *primary* sources consists solely of a mixture of hydrogen and helium (the stuff of zero-metallicity stars). The other *primary* source consists of the nuclear matter jetted out from the galactic center that yields not only the fissionable elements that ignite thermonuclear fusion reactions, but virtually all elements heavier than hydrogen and helium. *Secondarily*, over their lifetimes stars may synthesize some elements internally as well as possibly accumulating debris from previous astrophysical trauma-events.

Speculations on the Nature of the Universe: All attempts at this point in time to understand the nature of the Universe should properly be described as speculation, not science. But *"admirable speculation,"* to use Galileo's words [34], is nevertheless an important part of science, as it represents an attempt to begin to understand a scientific unknown.

In 1929, Hubble [301] noticed that the more distant a galaxy, the more its spectrum of light is shifted toward the red. Hubble adopted the interpretation of Slipher [302] for galactic-spectrum-shifts as being Doppler shifts in frequency caused by radial velocity. To Hubble and to those who followed, essentially all of the galaxies are moving away from us, and the further they are from us, the faster that they are moving away. So, how can that be? If the interpretation of Doppler shift is correct, which I seriously doubt, then it must mean that the Universe is expanding. That interpretation is the underlying basis for the *big bang* theory that the Universe is expanding from a point of nothingness. Nonsense!

The implicit assumption underlying Hubble-expansion is that if there were no expansion, light would travel *forever* without changing frequency and wavelength. Many astronomers, going back to Johannes Kepler (1571-1630), have noted, in their own ways that if the Universe is not expanding and is more-or-less homogeneous and is essentially *infinite* in size and has been in existence essentially *forever*, then the night sky should be filled with background light. But the night sky appears dark, simply lighted by points of light from stars and distant galaxies.

But behold! The sky is indeed filled with background light, but light not visible to the human eye. That light has lengthened in wavelength and is in fact the cosmic microwave background electromagnetic radiation discovered by Penzias and Wilson [303] (not a relic of the *big bang*, as some believe). Light, I posit, lengthens in wavelength on its long transit through interstellar space as it loses energy/mass through interaction with the *infinitesimal* matter along its sojourn, thus redistributing its energy/mass throughout a portion of the Universe, approaching cosmic equilibrium between its electromagnetic radiation and the *infinitesimal* matter.

Presumably, with much latitude for speculation, that *infinitesimal* matter becomes, through yet unknown reactions, hydrogen and helium, the primordial elements that in turn become the stuff of dark stars, which then gravitationally attract and form dark galaxies. As the galactic dark matter coalesces and becomes extremely dense at its center, at some point it begins to shoot out galactic jets. These jets, consisting of the parent nuclear matter for elements heavier than hydrogen and helium, seed dark stars they contact with fissionable elements and produce nuclear fission chain reactions, thus providing the million-degree temperatures necessary to ignite thermonuclear fusion reactions that light the formally dark stars. The now luminous stars radiate their visible light out into the

Universe, beginning anew the redistribution of energy/matter in the Universe.

Thus the Universe has no obvious beginning, and no foreseeable end. Presumably, the Universe is finite, yet unbounded.

Philosophical Implications: That the Universe has no obvious beginning, and no foreseeable end has both philosophical and theological implications. In the cosmology described here, the beginning, end, and age of the Universe are no longer ascertainable by scientific methodologies. In this instance, science no longer trumps theology.

Critical knowledge about the thermonuclear ignition of stars, and implications derived therefrom, although published in the scientific literature, has been un-cited and ignored by the Deep State scientific community. This is indicative of a bigger and far more devastating problem. Blatant deception and failure to tell the truth pervades officialdom worldwide and should not be tolerated, as these practices pose very real threats to civilization, and to individual freedom.

5 DEEP STATE ENVIRONMENT DESTRUCTION AND DECEIT

Anyone with a deep connection to nature can see how badly the natural world is suffering. Gone are the days of lush green forests and hillsides, clear blue skies, and star-studded nights with the Milky Way Galaxy clearly visible. Fields and roadsides no longer bustle with insect life and each spring and fall migration brings fewer and fewer birds. An overturned rock that once revealed an entire community of living organisms is now barren beneath. The summer heat has become unbearable, and you can feel the burn of the sun on your skin. Forests are disappearing and remaining trees display thinned-out foliage with trunks and branches scorched and damaged by the sun and runaway fires. Coral reefs are dying everywhere, and the oceans are grossly polluted and brimming with harmful algae blooms. There is widespread desertification of the land and brownification of the world's surface waters. Anyone looking up can see the awful chemical trails fanning out to create a milky, sun-blotted sky must realize that humanity has made a real mess of things and we are in serious trouble.

Elders among us may remember with fondness bygone images of unmarred nature. The richness and diversity of life on Earth is disappearing at an incredible rate. Beyond the explosion of species extinction, there are massive population declines of both plants and animals with cascading effects on ecosystems necessary for our continued existence [304]. Human activities have destroyed over two-thirds of the world's wildlife in just the past fifty years [305,

306] and there is no end in sight. Few scientists have found the courage to sound the alarm about our dire situation [307] and fewer among those realize that much of our current environmental crisis is deliberately caused.

In 1962, Rachel Carson [308] published *Silent Spring* to alert the public of the devastating environmental effects of the pesticide DDT and other toxic chemicals whose manufacturers were lying about their safety. *Silent Spring*, widely read in part as the result of being serialized by *The New Yorker* and selected for Book of the Month in October, began to open people's eyes to the harm these chemicals pose to the environment. Concerned citizens became environmental activists, some forming organizations to educate, politicize, and litigate for a clean, healthy environment.

For decades, environmental organizations garnered large sums of money from the public, as well as from foundations and government agencies. Now, in 2025, not only is the environment immeasurably worse than it was in 1962, but we have already entered an unmistakable phase of biosphere collapse. The herbicide glyphosate [309] is the DDT of the 21st century, and that is not all. Unbeknownst to all but a few enlightened citizens, deliberate environmental degradation and ruination have become covertly institutionalized on an international level. Even when brought to the attention of mainstream media, elected officials, government agencies, national scientific academies, environmental organizations, medical publishers – to name a few – these organizations refuse to acknowledge any evidence of the intentional assault on our global environment. Worse, there are deliberate attempts to deceive the public of the concomitant health risks posed by such deliberate, ruinous activity [310, 311].

In the same decade as publication of *Silent Spring*, humans for the first time ventured into space and viewed Earth as an island in the great void of outer space. From that view sprang awareness that ours is the only planet in the solar system capable of sustaining human life. In the early 1970s, James Lovelock and Lynn Margulis [312-314] co-authored the *Gaia hypothesis* likening Earth to a biome made up of a vast number of interconnected and interacting physical and biologically based processes.

Life on Earth is perilous even without adverse human activity. In the geologic past there have been at least five major episodes of large-scale species extinction [162]. Natural disasters alone pose great risks to humanity, including earthquakes, volcanic eruptions, wildfires, storms, and changes in the geomagnetic field that potentially increase threats from the onslaught of the solar wind. Nevertheless, interactions by and between myriad natural physical and biological processes have made life possible on this planet for several billion years. This complex and delicate balance in nature is now under threat from human intervention that, if unchecked, will almost certainly lead to the extinction of *Homo sapiens*.

The challenge for humanity is two-fold: At one level, humans need to become good stewards of their only home planet, living and working in ways that are harmonious with and non-toxic to the natural environment. Much has been written on this subject, but little progress is being made [315-317]. At another level, humans need to become aware of and put a stop to deliberate, covert actions that are aimed at destroying the natural environment for political and economic gain.

At least as far back as the 1990s, concerned citizens had started to notice white trails that stretched across and dimmed the sky, diffusing briefly to resemble cirrus clouds before making a white haze. As time passed, these jet-laid white trails became more frequent and were observed over wider geographical regions. By 2012 these trails had become a near-daily, near-global occurrence (Figure 5.1).

Figure 5.1. Coal fly ash jet-laid trails with photographers' permission, from [318]. Clockwise from upper left: Soddy-Daisy, Tennessee, USA; Reiat, Switzerland; Warrington, Cheshire, UK; Alderney, UK looking toward France; Luxembourg; New York, New York, USA.

Concerned citizens had many questions: What substances were being jet-sprayed to form the trails? Why was this being done? What were the risks to human and environmental health? What was the legal justification? Inquiries to authorities inevitably were met with the same response, namely, these are *contrails*, harmless ice crystals from the moisture in jet's exhaust [319]. Nonsense! The jet-laid trails, which some refer to as *chemtrails*, behave quite differently than contrails [320].

Ice-crystal contrails tend to persist for more than a few seconds (or sometimes minutes) only when there is considerable moisture in the aircraft exhaust and the ambient atmosphere is both cold and humid. Commonly, especially with modern fan jet engines, ice crystals from exhaust evaporate quite quickly becoming invisible gas, not remaining as a haze to scatter sunlight. Additional observations and scientific measurements demonstrate quite conclusively that chemtrails are *not* contrails [321].

Many concerned citizens wanted answers about the chemtrail phenomenon, and realized that they were getting disinformation or no information at all from various authorities. It was a major concern that the scientific and medical communities in the United States, British Commonwealth nations, and the European Union turned a blind eye to the deliberate jet-emplaced air pollution. Clearly, something was badly amiss and potentially devastating on a global scale.

Meantime, some concerned citizens had taken samples of post-spray rainwater, and had the samples analyzed at commercial laboratories and posted the results on the Internet. Most individuals only requested aluminum analyses, some also requested barium, while a few requested analyses of aluminum, barium, and strontium. In response to the findings, one widespread published explanation, presumably made to mislead or deceive people, is that chemtrails consist of the oxides and/or sulfates of aluminum, barium, and strontium, which are relatively harmless substances because they are practically insoluble in water. That explanation, however, is contrary to peer-reviewed analytical data on chemtrails that shows aluminum, barium, and strontium are in fact *dissolved* in rainwater!

Some unknown powdered substance, covertly mixed in jet-fuel [322] and/or jet-sprayed into the lower atmosphere (troposphere), produces chemtrails. This substance reacts with moisture, causing some of its chemical elements to be extracted into atmospheric water. But what substance? Sprayed by whom? And why? Clearly, forensic science – scientific detective work – was needed.

It soon became clear that many million metric tons of this unknown substance were being jet-sprayed into the atmosphere annually, but there were no obvious sources for such massive quantities of aerosol pollutants, no conspicuous production facilities. Yet somewhere massive-scale production had to be taking place. The unknown substance had to contain aluminum – found repeatedly in all rainwater samples taken – and was clearly not a natural product, such as desert sand, because the Earth's surface aluminum is generally chemically combined, locked up tightly with oxygen and does not dissolve in rainwater.

Toxic Coal Fly Ash: Throughout the academic scientific literature there are numerous references to a toxic waste produced by industrial coal burning called *coal fly ash* [323-325]. The annual global production of coal fly ash reported in 2014 was 130 million metric tons [326]. This could be a sufficient supply to jet-spray as an aerosol on the scale observed. Notably, a Spanish laboratory's experiments were conducted and published in 2005, mixing coal fly ash with distilled water for 24 hours. The results showed that at least 38 elements were partially dissolved in the water [327]. The dissolved elements included aluminum, barium, and strontium.

In 2015, Indian scientists sought assistance to understand the geological association of highly mobile aluminum with human health in the Alluvial Plain of the Ganges River [328]. I responded [329], showing that Internet-posted pairs of elements, aluminum/barium and strontium/barium measured in rainwater were similar to corresponding element pairs extracted into water from coal fly ash from experiments conducted by the Spanish scientists [327]. Published in 2015, my *Current Science* paper was the first article in the scientific literature that not only mentioned "chemtrails" but provided the first scientific evidence that toxic coal fly ash was the main substance being jet-sprayed into the lower atmosphere (troposphere) to produce chemtrails.

Perhaps the best evidence of the *Current Science* paper's correctness was the immediate demand for retraction delivered not only to that journal but soon thereafter to two public health journals that also had peer-reviewed and published subsequent work by me on the adverse health consequences of deliberately aerosolized coal fly ash. In the latter instances, the US editors retracted the papers *without allowing the author to see or respond to the complaints* [330]. Such behavior is outrageous. Only about 1 in 15,000 peer-reviewed, published scientific articles is retracted, and that only occurs after the authors are confronted with the allegations of wrongdoing and are given the opportunity to defend themselves.

The concerted efforts to suppress publications warning of the public health risks of jet-sprayed coal fly ash was a clear indication that those who order or otherwise participate in the jet-spraying operations know the health risks involved and want to hide them from the public.

This was just the beginning. Much more scientific detective work had to be done. It was at this point that Mark Whiteside, M.D., M.P.H, the Medical Director of the Florida Public Health Service in Key West contacted me. We began a collaboration to bring together scientific and medical knowledge in a more comprehensive forensic investigation and framework. Along with occasional associates, we published more than 35 scientific and medical investigations and two books [331, 332]. The discoveries we made provide compelling evidence of the wholesale pollution of our atmosphere with coal fly ash and show how it alters our planet's physical environment to the detriment of all life, including the devastating effects on virtually all lifeforms, including humans. Moreover, we discovered the greatest secret of all, the so-called "legal" pretext, and true intent of the environmental modification.

Not content with using Internet-posted rainwater analyses, we personally collected or arranged for the collection and commercial analyses of post-spraying samples of rainwater and snow. The analytical results for 10 element pairs are shown in Figure 5.2 along with the previously mentioned Internet-posted pairs.

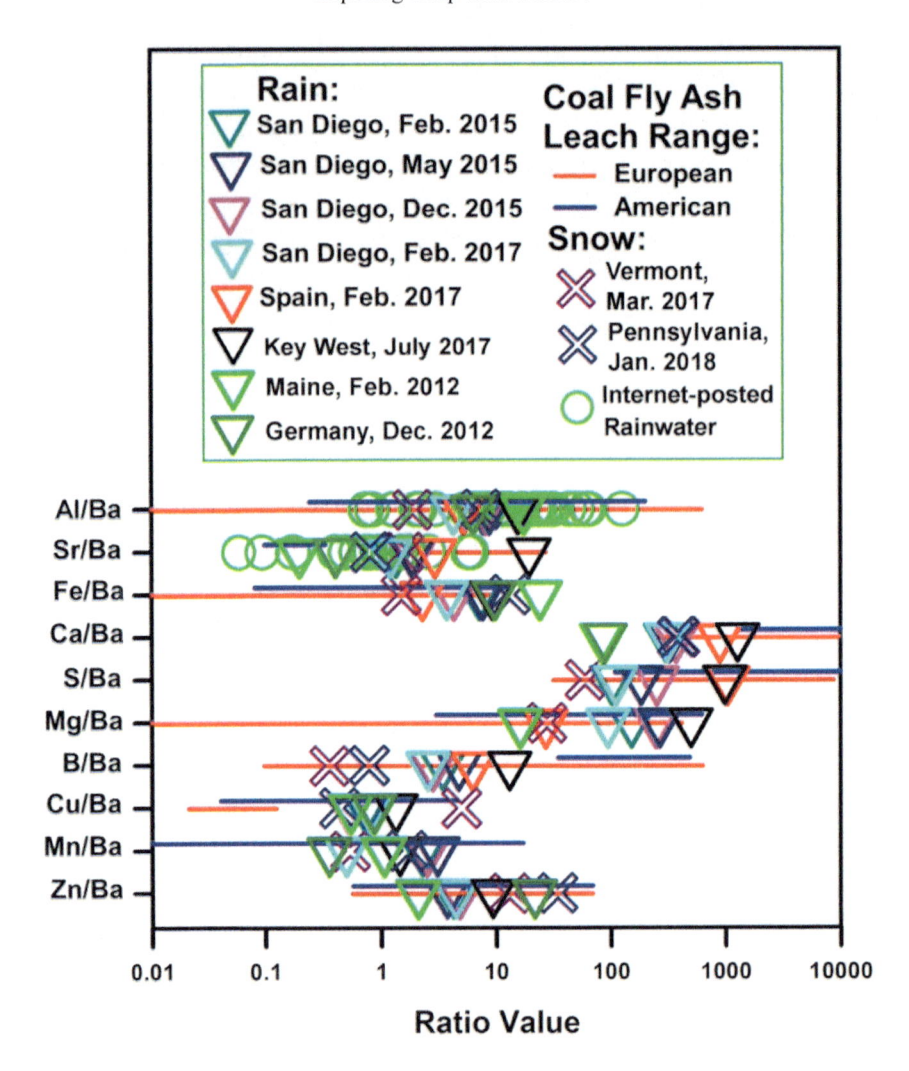

Figure 5.2. From [333], showing the similarity of element ratios measured in rainwater and snow with the range of comparable element ratios measured in the laboratory lixiviate of water-leach experiments [327, 334].

Falling snowflakes trap and bring down particles jet-sprayed into the lower atmosphere. Taking a sample of the snow, allowing it to melt and then to evaporate, left a residue that could be analyzed and compared to the range of values measured in various samples of coal fly ash (Figure 5.3).

In areas, such as northern USA and Canada, snow mold sometimes grows atop grass beneath snow. As the snow starts to melt, the particles it trapped are released and may be trapped again on the underlying snow mold. Figure 5.3 also shows analytical values for element pairs from snow mold and for particles dropped from an aircraft and collected where they fell on an automobile in Encinitas, California (USA).

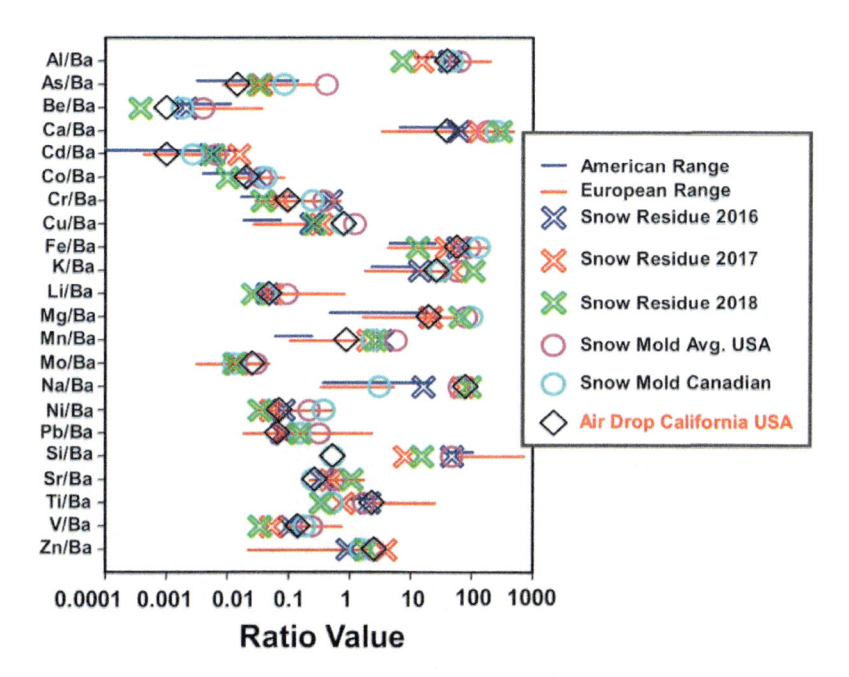

Figure 5.3. From [335], comparison of analytical results with the ranges of European [327] and American [334] coal fly ash samples.

Air pollution, the world's leading cause of environmental human fatalities, is a major contributor to noncommunicable disease. Aerosolized coal fly ash, a particularly hazardous form of air pollution, pours forth from numerous smokestacks in India and China. However, citizens of the United States, the British Commonwealth, and the European Union live in a fools' paradise. They have been led to believe that coal-burning utilities in their countries trap this very toxic substance so it does not exit smokestacks and directly pollute the air. They do store the exhaust-precipitates as solid waste, but then the utilities surreptitiously supply the coal fly ash to be covertly jet-sprayed into the very air people breathe, and indeed profit from that diabolical activity.

Coal fly ash has thus become a toxic environmental nightmare consisting mainly of tiny spherical particles (Figure 5.4). These particles contain concentrations of the most hazardous chemical elements in coal, each of which can harm the natural environment in numerous ways. For example: Aerosolized coal fly ash contaminates the environment with mercury, one of the most toxic poisons known, and known to move up the food chain [336]. Aerosolized coal fly ash, lofted into the upper atmosphere (stratosphere), destroys Earth's ozone layer [337-339], and exposes all surface life to the deadly ultraviolet radiation from the sun [340]. Contaminating the environment with the massive quantities of iron contained in coal fly ash upsets the delicate iron balance in nature and in the bodies of exposed biota [341, 342].

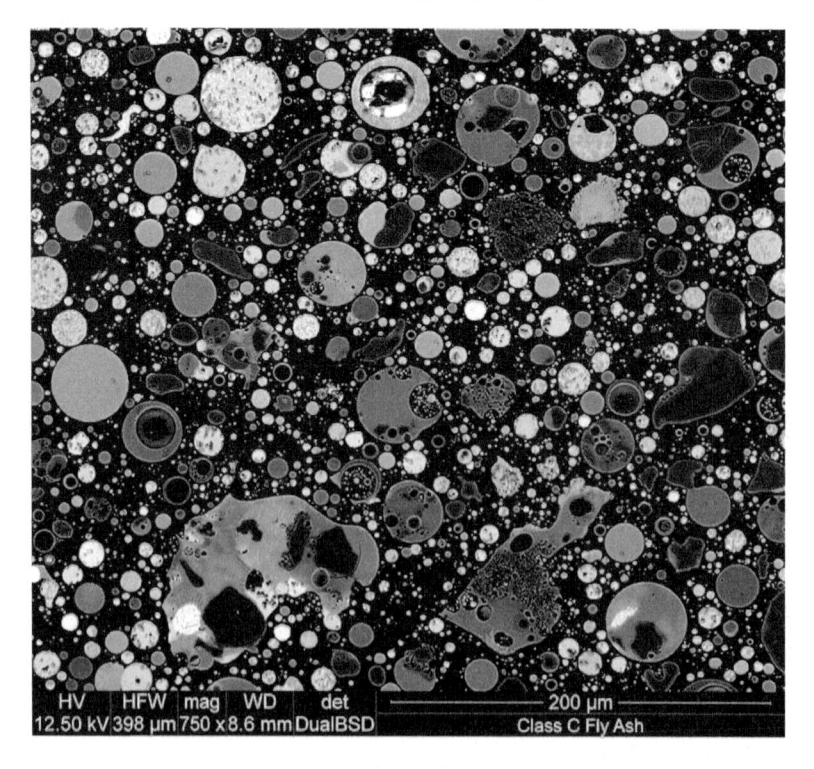

Figure 5.4. Polished section of coal fly ash embedded in epoxy. From [320].

Aerosolized coal fly ash's ultrafine particles and nanoparticles can enter the blood stream through the nasal bulb or through alveoli in the lungs. These particles can collect in the brain [343, 344] and in the heart [342]. When exposed to body fluids, coal fly ash can release a host of toxic chemicals including neuro-toxic, chemically mobile aluminum, and carcinogens such as arsenic, hexavalent chromium, as well as the ash's radioactive elements. The elements from coal fly ash may produce many toxic effects, including decreased host defenses, tissue inflammation, altered cellular redox balance toward oxidation and genotoxicity which can lead to chronic lung disease [345], lung cancer [346], and neurodegenerative diseases [343].

Particulate pollution has multiple, serious adverse consequences on environmental health. Several lines of evidence now point to particulate pollution as a possible co-factor in the COVID-19 Pandemic, specifically, as a potential means of viral transport, as a co-factor in exacerbating susceptibility and mortality, and in diminishing immune response to the SARS-CoV-2 virus [347, 348]. The association of particulate pollution with the COVID-19 Pandemic is a wake-up call to humanity that foreshadows *even greater global devastation through undisclosed particulate pollution.*

Humans are not the only life form adversely affected by jet-emplaced coal fly ash. Plants, trees and indeed whole forests are harmed by coal fly ash chemtrails in three main ways [349, 350] (Figure 5.5).

Figure 5.5. Dead Torrey Pines silhouetted against a sky displaying jet-emplaced particulate trails. From [349].

Plants and trees are damaged by chemtrail-caused drought and poisoned by the chemically mobile aluminum added to atmospheric water. They are also damaged by the increased levels of harmful solar ultraviolet radiation caused by chemtrail-destruction of ozone, which shields life against dangerous ultraviolet radiation. Orchards and agricultural plants are affected similarly [351].

Atmospheric manipulation that utilizes aerosolized coal fly ash is a primary factor in the extent and severity of forest fires in California and elsewhere; other adverse effects include exacerbation of drought, tree and vegetation desiccation and die-off, and the unnatural heating of Earth's atmosphere and surface regions [320]. Forest combustibility is increased by moisture-absorbing aerosolized particles that damage the waxy coatings of leaves and needles, reducing their tolerance to drought. The aerosolized coal-fly-ash-chemtrail climate manipulation greatly increases the potential for forest fire ignition by lightening. Wildfires dramatically worsen baseline air pollution, emitting harmful gases and volatile organic compounds, and they both concentrate and re-emit toxic elements and radioactive nuclides over wide areas. The type of air pollution created by wildfires is associated with increased all-cause mortality worldwide, with the greatest impact on respiratory and cardiovascular disease.

Wildlife species are suffering a precipitous global decline [304]. Aerosolized coal fly ash is a significant factor in the catastrophic global decline in the populations of insects [352], birds [335], and bats [353]. Insects can ingest and/or accumulate toxic coal fly ash on their body surfaces which then birds and insectivorous bats consume. Coal-fly-ash-chemtrails disrupt the natural environment, modify habitats, and adversely affect the natural life cycles, upsetting the complex and delicate balance that makes life possible on Earth. Even marine environments are affected; for example, chemtrails cause shifts in the global plankton community balance in favor of harmful algae and cyanobacterial blooms in fresh and salt water [341] (Figure 5.6). And corals are harmed by increased levels of solar ultraviolet radiation [354].

Figure 5.6. *Karenia brevis* **and** *Trichodesmium erythraeum* **Bloom, Offshore Lee County, Florida (USA), October 22, 2007. Photo courtesy of Florida Fish and Wildlife Conservation Commission. From [341].**

In 1968, MacDonald [355] wrote an influential essay on environmental warfare entitled "How to Wreck the Environment" which, among other revelations included his thoughts on weather warfare: "...*removing moisture from the atmosphere so that a nation dependent on water...could be subjected to years of drought. The operation could be concealed by the statistical irregularity of the atmosphere. A nation possessing superior technology in environmental manipulation could damage an adversary without revealing its intent ... one can conduct covert operations using a new technology in a democracy without the knowledge of the people.*"

Covertly disrupting natural weather can lead to crop failures, livestock decimation, and starvation; all of which can destabilize political regimes and exacerbate potential hostilities [351]. Indeed, former Iranian President Mahmoud Ahmadinejad accused Western nations of surreptitiously causing droughts in his country [356].

Figure 5.7 shows particulate trails blanketing the Republic of Cyprus whose citizens, so far unsuccessfully, sought an explanation from their government for the deliberate obscuration of their skies [357].

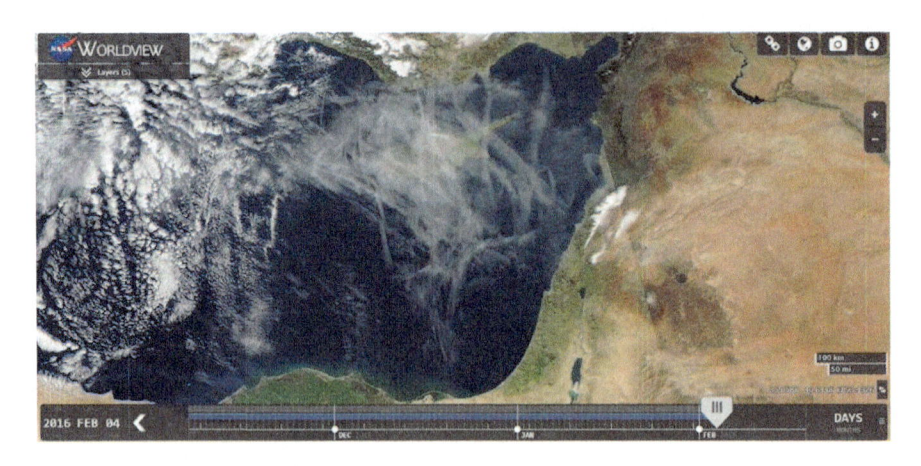

Figure 5.7. NASA Worldview satellite image from February 4, 2016 showing jet-laid trails blanketing the air above the Republic of Cyprus but nearly absent in surrounding regions. From [358].

Coal fly ash jet-emplaced into the troposphere, in addition to suppressing rainfall, serves other purposes. For example, chemical elements that are extracted from coal fly ash into atmospheric moisture substantially increases its electrical conductivity [327], making it possible to move weather masses with electromagnetic radiation (Figure 5.8). Although cloaked in great secrecy, conceivably it is now possible to weaponized hurricanes and cyclones by altering their paths.

Figure 5.8. Example of electromagnetic radiation being used for weather manipulation. From [359].

While it is easy to imagine that major militaries would delight in having the means to develop the know-how to weaponized weather, clearly there is a much bigger agenda involved; there are too many circumstances that seem to defy reason. For example, the conspicuous jet-laid particulate trails occur over many different countries which, except in very rare instances, do not object or levy blame. Moreover, a pervasive *omertá*, code of silence, seems to exist internationally, which is strange as the aerial particulate emplacement has devastating human and environmental health consequences.

Why? As we discovered the multiple, ramifying consequences of chemtrail coal fly ash on the environment, and on human and environmental health, one nagging question remained: Why? Why was the spraying occurring worldwide without let-up year after year?

Humans are opportunistic creatures. For example, many, lacking a moral compass, might capitalize on the consequences of the chemtrail assault on the environment by acquiring farmland from farmers whose crops were ruined by chemtrails, or real estate destroyed by rampant forest fires. These are merely examples of consequences, not their cause. Why would supposedly civilized people keep silent and allow the wholesale destruction of human and environmental health? Why is this happening? What is the supposedly legal basis? And, what entities are responsible? These questions also shaped our forensic scientific and medical investigation.

In 1968, in the same decade *Silent Spring* was published, geophysicist Gordon J. F. MacDonald published a book chapter entitled "How to Wreck the Environment" [360] in which he described how one nation might trigger the forces of nature to wage environmental warfare on an enemy nation. Fifty years later we reviewed that work in light of technological advances made since it was written [358].

The aerial chemtrail assault with all its secrecy and disinformation now began to make perverse sense: It is a form of environmental warfare, seemingly waged on behalf of a future globalized civilization, a new civilization to be born of the ashes of a new form of world war. We discovered the clever, but deceitful and deceptive 1978 United Nations International Treaty (ENMOD), a *Trojan horse* legal agreement designed to cause sovereign nations to wage covert environmental warfare against their own citizens and against citizens of other sovereign nations [361-363].

Figure 5.9 shows the global distribution of ENMOD signatories.

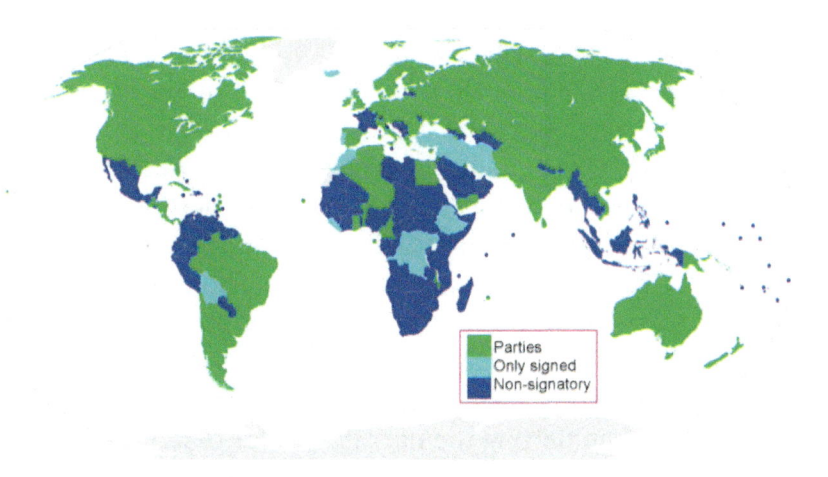

Figure 5.9. A public domain image image showing the Janusry 3, 2018 distribution of ENMOD signors. Adapted from [362].

As we revealed [362, 364], the 1978 *"Convention on the Prohibition of Military or Any Other Hostile Use of Environmental Modification Techniques"* (ENMOD) obligates signatory nations to *fundamentally compromise their own sovereignty* and to bring about widespread, permanent agricultural devastation. Instead of *prohibiting* "Hostile Use of Environmental Modification Techniques", as its title suggests, ENMOD *obligates* signatory nations to participate in unspecified "peaceful" environmental modification activities performed by unspecified entities, under unspecified circumstances, without limitation to harm. Whether harm is inflicted on a nation or a region's agriculture, its environment, or on the health of its citizenry does not matter from ENMOD's international legal point of view because its intent is "peaceful."

Regardless, large-scale *environment modification* cannot be construed as "peaceful." It is instead fundamentally hostile.

The veil of ENMOD deception was pierced by applying precise knowledge of contract law to ENMOD's Articles [362]. The highly secret *"peaceful"* environmental modification project activity was discovered by an accidental release of material from an aircraft in 2016 [365, 366].

Further proof that global warming is an environmental modification (geoengineering) goal was provided on or about February 14, 2016, when an oily-ashy substance was accidently released by an aircraft and fell on seven residences and vehicles in Harrison Township, Michigan (USA). Upon reading the news reports, I contacted one of the home owners who provided samples and photographs. I found the "air-drop material" consists of an assemblage of plant material mixed with coal fly ash and salt. The splatter pattern on the vehicles, ground, and rooftops in Harrison Township resembled the cryoconite holes observed on ablating glaciers. The "air-drop material" appeared to have been modeled after natural cryoconite, with coal fly ash imparting the dark gray color that absorbs sunlight, melts glacial ice, and contributes to global warming [365-367] (Figure 5.10).

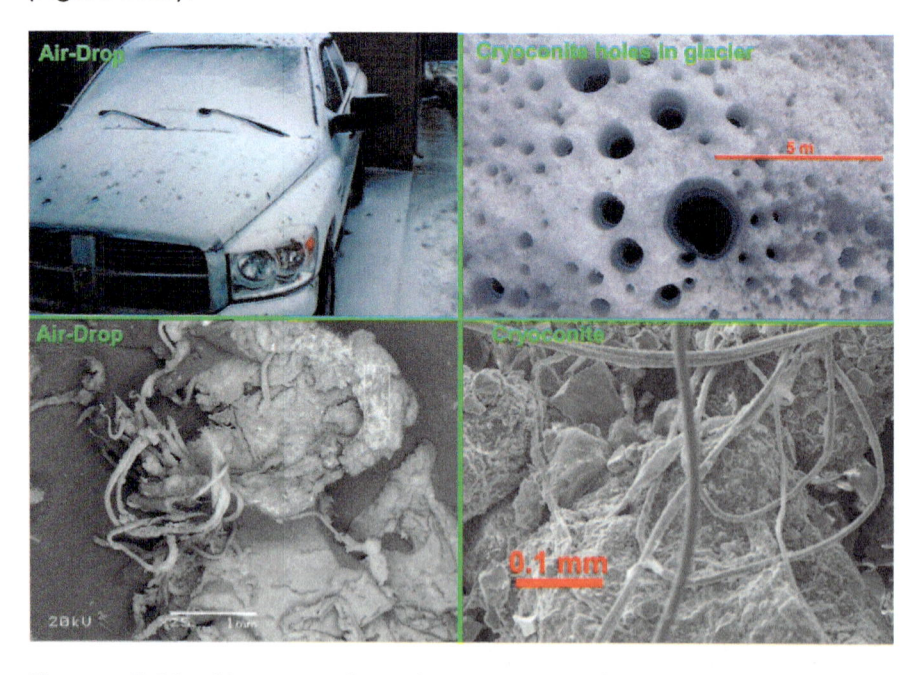

Figure 5.10. Upper Left: Air-Drop Distribution; Upper Right: Cryoconite-hole Distribution in Glacier; Lower Left: Air-Drop Synthetic Cryoconite; Lower Right: Natural Cryoconite. From [366].

We regard it a near certainty that the highly secret ENMOD "peaceful environmental improvement," sanctioned and, indeed, even mandated by the treaty, is to melt Arctic ice to open sea lanes above the Arctic Circle.

Altering Earth's natural environment on such a scale is not "peaceful," whether the goals are commercial or military. The true nature of this United Nations ENMOD sanctioned activity, we allege, is the conduct of covert environmental warfare against established nation states, including Western states, their citizens and their culture in favor of a supranational, atheistic, global regulatory regime whose oversight will encompass a fully globalized economy dominated by multinational mega-corporations [361-363].

ENMOD's purpose is to subvert the authority of individual nation states.

What other evidence points to direct United Nations involvement?

United Nations Global Warming Mistakes and Lies: Climate change, sometimes called global warming, is currently a high-profile subject of public debate, and a matter of grave concern for government leaders and educators. Those individuals ought to be able to rely on trained scientists for guidance, for true understanding, but there is a problem. Science has become tainted by politics. That was not always the case.

In the highly politicized domain of *climate change*, there are two schools of thought: The majority view is that global warming is occurring, caused by carbon dioxide (CO_2) and other greenhouse gases trapping heat that would otherwise be released into space. This majority viewpoint is mainly the result of United Nations' propaganda. The minority view is that there is no global warming, and any variation in global temperature is natural. This minority viewpoint is mainly the result of oil industry propaganda. Paradigm changing scientific understanding, as I have published [15, 318, 367-371] and described below, clearly leads to the conclusion that neither school of thought is correct: Global warming is in fact occurring, but not primarily as a consequence of greenhouse gases. Global warming is primarily caused by atmospheric aerosol particles [15, 318, 367-371].

The good news is that reducing particulate pollution will rapidly reduce global warming in a matter of days to weeks.

For more than three billion years, as long as life has existed on Earth, the surface of our planet has maintained a remarkably stable state of thermal equilibrium through the combined-effect of numerous natural processes, despite being bombarded by variable solar radiation from above [372, 373] and variable planetary energy sources from below, including georeactor nuclear fission energy [10, 13] and stored protoplanetary compression energy [12, 122, 123]. Decades ago, considering the ever-increasing scale of human activity, it might have been prudent to engage in open scientific debates and discussions to ascertain with reasonable certainty the nature and extent that human activities might be altering those natural processes. But, to my knowledge, such objective, open inquiry never occurred.

Instead, in 1988 the United Nations' Intergovernmental Panel on Climate Change (IPCC) was established, and in concert with various other political entities, presumably driven by political and/or financial motives [255], convinced numerous government officials that greenhouse gases, notably fossil-fuel produced carbon dioxide (CO_2), were trapping heat that otherwise should have been released to space [374]. Climate change, also known as global warming, became a new global enemy.

The science promulgated by the IPCC and the climate science community is badly flawed [369]. Some factors that affect climate are poorly known, some others are misrepresented or ignored [375]. The most serious flaw, however, is the wide-spread use of *"climate models"*, assumption-based computer programs which typically begin with a known end-result which they attain by cherry-picking data and parameters [113]. Climate models, being computer programs, are subject to the well-known dictum *"garbage in, garbage out"* [376].

Climate scientists rarely acknowledge how much remains unknown, outside the bounds of their computational models. Exceptionally, Curry and Webster [377] state: *"In addition to insufficient understanding of the system, uncertainties in model structural form are introduced as a pragmatic compromise between numerical stability and fidelity to the underlying theories, credibility of results, and available computational resources."*

As James Lockwood noted [378]: *"Gradually the world of science has evolved to the dangerous point where model-building has precedence over observation and measurement, especially in Earth and life sciences. In certain ways modeling by scientists has become a threat to the foundation on which science has stood: the acceptance that nature is always the final arbiter and that a hypothesis must always be tested by experiment and observation in the real world."*

The climate science community treats global warming solely as a radiation-balance issue, based upon the idea that all of the heat received from the sun, as well as the heat brought to the surface from deep-Earth heat-sources, must be released to space. In that paradigm, they define an artificial construct "radiative forcing" or "climate forcing" in units of Wm^{-2} relative to 1750 Wm^{-2} as a means to represent the departure from zero-net radiation balance [379], assumed to be caused primarily by anthropogenic carbon dioxide and other greenhouse gases. While that approach is convenient for computer model results, it leads to a false understanding of the factors affecting Earth's surface temperature.

Aerosol Ignorance: The United Nations' Intergovernmental Panel on Climate Change (IPCC) and the climate science community generally subscribe to the false proposition that tropospheric aerosol particulates cool the climate [377, 380, 381], with the exception of black carbon aerosols [382]. IPCC scientists maintain that the consequence of aerosolized particulates is to block sunlight and cool the Earth [380, 383-385].

Climate scientists undervalue the role of aerosols and clouds in trapping heat, contending that heat trapping occurs primarily by atmospheric greenhouse gases as evidenced by the following statement [380]: *"Atmospheric aerosols counteract the warming effects of anthropogenic greenhouse gases by an uncertain, but potentially large, amount....Strong aerosol cooling in the past and present would then imply that future global warming [due to pollution reduction] may proceed at or even above the upper extreme of the range projected by the Intergovernmental Panel on Climate Change."* The perception of tropospheric aerosol particulates' cooling effect on Earth's climate has led to fundamental misconceptions in climate science.

Many climate scientists falsely believe that aerosol-particles, including black carbon, cool the Earth's surface [377, 380, 381, 383-388] or are uncertain whether aerosols cool or heat the Earth [389, 390]. For example, Ramanathan and Carmichael [391] state: *"...black carbon has opposing effects of adding energy to the atmosphere and reducing it at the surface."* Similarly, Andreae, Jones and Cox [380] state: *"Atmospheric aerosols counteract the warming effects of anthropogenic greenhouse gases by an uncertain, but potentially large, amount."* Uncertainty as to whether aerosols result in cooling or warming hinders the ability to predict future climate changes [392, 393] and even hinders the ability to understand the factors that are responsible for maintaining surface temperatures in a range that makes life possible.

As noted in Chapter 1, science progresses by questioning the correctness of popular paradigms and through tedious efforts to place seemingly independent observations into a logical order in the mind so that causal relationships become evident and new understanding emerges [36]. Rather than making grand, detailed, computational-models based upon poorly understood complexities of climate science, a much better way would be to better understand the behavior of specific factors that affect Earth's climate.

Evidence from World War II: Harvard physicist Bernard Gottschalk [394, 395] noticed a thermal peak coincident with World War II (WW2) in a global temperature profile image on the front page of the January 19, 2017 *New York Times*, and was inspired to investigate further. He applied sophisticated curve-fitting techniques to eight independent global temperature datasets from the US National Oceanic and Atmospheric Administration (NOAA), demonstrated that the WW2 peak is a robust feature, and concluded that the thermal peak *"is a consequence of human activity during WW2"* [394, 395].

The conspicuous aspect of Gottschalk's global-warming curves [394], shown by the black curves in Figure 5.11, is that immediately after WW2 the global warming rapidly subsided. That behavior is inconsistent with CO_2-caused global warming as CO_2 persists in the atmosphere for decades [396, 397]. Furthermore, CO_2-caused global warming during WW2 can be ruled out as Antarctic Law Dome Ice core data during the period 1936-1952 show no significant increase in CO_2 during the war years, 1939-1945 [398].

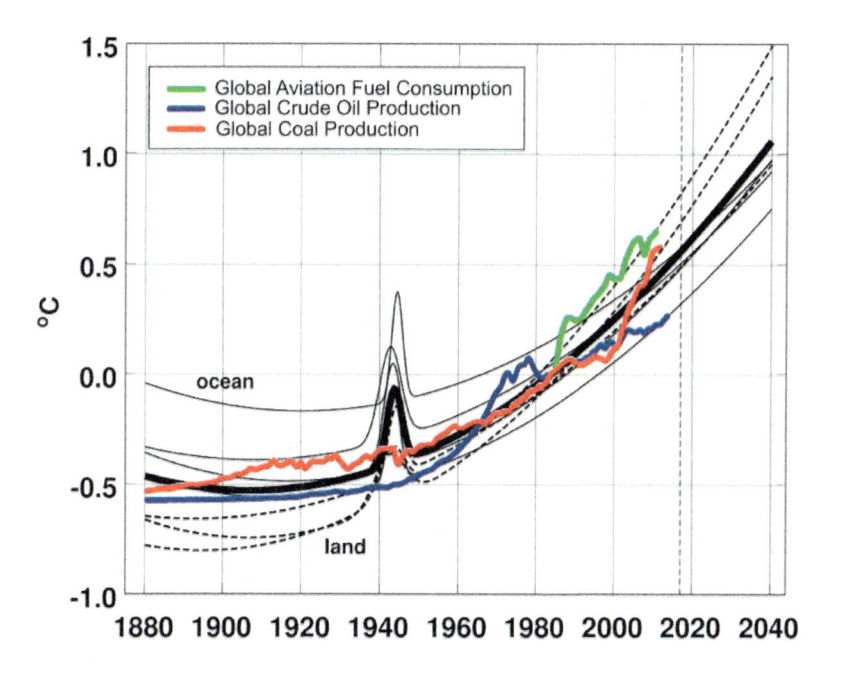

Figure 5.11. Copy of Gottschalk's fitted curves for eight NOAA data sets showing relative temperature profiles over time [394] to which I added proxies for particulate pollution. Dashed line: land; light line: ocean; bold line: weighted average. From [15].

I realized a different explanation. World War II activities injected massive amounts of particulate matter into the lower atmosphere (troposphere) from extensive military industrialization and vast munition detonations, which included demolition of entire cities, and their resulting debris and smoke. The implication is that the aerosolized pollution particles trapped heat that otherwise should have been returned to space, and thus caused global warming at Earth's surface [15] which would have subsided rapidly after hostilities ceased. Rapid cessation of WW2 global warming is understandable as tropospheric pollution-particulates typically fall to ground in days to weeks [399-403].

Figure 5.11, from [15, 394], shows relative-value, particulate-pollution proxies added to Gottschalk's figure: Global coal production [404, 405]; global crude oil production [405, 406]; and, global aviation fuel consumption [405]. Each proxy dataset was normalized to its value at the date 1986, and anchored at 1986 to Gottschalk's boldface, weighted average, relative global warming curve. The particulate-proxies track well with the eight NOAA global datasets used by Gottschalk [15].

Following the end of WW2 hostilities, wartime aerosol particulates rapidly settled to ground [399], Earth radiated its excess trapped energy, and global warming abruptly subsided for a brief time [15]. Soon, however, post-WW2 industrial growth, initially in Europe and Japan, and later in China, India, and the rest of Asia [407] increased worldwide aerosol particulate pollution and with it concomitant global warming [15].

From the evidence shown in Figure 5.11, there is one inescapable conclusion: *Aerosol particulate pollution, not carbon dioxide, is the main cause of global warming.* That conclusion was not at all evident from the "radiation-balance" methodology and parametrized models widely utilized by the climate science community. The concept that aerosol particulate pollution is the main cause of global warming thus constitutes a climate-science paradigm shift. Yet, in certain respects it might have been obvious to those who observe nature and who reason objectively. For example, one might have observed that in the desert cloudy days are usually cooler than non-cloudy days, while cloudy nights are typically warmer than non-cloudy nights.

Global Warming by Aerosol Particulates: Most particles found in the troposphere absorb solar energy to some extent from one or more portions of the wavelength spectrum [408-414]. As Hunt noted [415]: *"A dispersion of small absorbing particles forms an ideal system to collect radiant energy, transform it to heat, and efficiently transfer the heat to a surrounding fluid.... If the characteristic absorption length for light passing through the material comprising the particles is greater than the particle diameter, the entire volume of the particles is active as the absorber. When the particles have absorbed the sunlight and their temperature begins to rise they quickly give up this heat to the surrounding gas...."*

Aerosol particles are heated by solar radiation and by radiant heat from the Earth, and transfer that heat to atmospheric gases by molecular collisions. The resultant atmospheric heating has the consequence of reducing atmospheric convection and thus reducing heat loss from Earth's surface [367, 370].

Convection is perhaps the most misunderstood natural process in Earth science. Hypothetical convection models of Earth's fluid core [416-420] and of Earth's mantle [421-423] continue to be produced, although sustained thermal convection in each instance has been shown to be physically impossible [10] thus necessitating a fundamentally different geoscience paradigm [8, 9, 12, 13, 122, 123, 424].

Chandrasekhar described convection in the following, easy-to-understand way [251]: *"The simplest example of thermally induced convection arises when a horizontal layer of fluid is heated from below and an adverse temperature gradient is maintained* [i.e. bottom hotter than top]. *The adjective 'adverse' is used to qualify the prevailing temperature gradient, since, on account of thermal expansion, the fluid at the bottom becomes lighter than the fluid at the top; and this is a top-heavy arrangement which is potentially unstable. Under these circumstances the fluid will try to redistribute itself to redress this weakness in its arrangement. This is how thermal convection originates: It represents the efforts of the fluid to restore to itself some degree of stability."*

To the best of my knowledge, consequences of the *adverse temperature gradient*, described by Chandrasekhar [251] have not been explicitly considered in either solid-Earth or tropospheric convection calculations. The following simple classroom-demonstration experiment, however, can provide critical insight for understanding how convection works that is applicable to both tropospheric and Earth-core convection [370].

As described recently [367]: *The convection classroom-demonstration experiment was conducted using a 4 liter beaked-beaker, nearly filled with distilled water to which celery seeds were added, and heated on a regulated hot plate. The celery seeds, dragged along by convective motions in the water, served as an indicator of convection. When stable convection was attained, a ceramic tile was placed atop the beaker to retard heat loss, thereby increasing the temperature at the top relative to that at the bottom, thus decreasing the adverse temperature gradient.*

Figure 5.12, from [370], extracted from the video record [425, 426], shows dramatic reduction in convection after placing the tile atop the beaker. In only 60 seconds the number of celery seeds in motion, driven by convection, decreased markedly, demonstrating the principle that reducing the adverse temperature gradient decreases convection. That result is reasonable as zero adverse temperature gradient by definition is zero thermal convection.

Figure 5.12. A beaked-beaker of water on a regulated hot plate with celery seeds pulled along by the fluid convection motions. Placing a ceramic tile atop the beaker a moment after T=0 reduced heat-loss, effectively warming the upper solution's temperature, thus lowering the adverse temperature gradient, and reducing convection, indicated by the decreased number of celery seeds in motion at T=60 seconds. From [370].

Particles in the troposphere, including the moisture droplets of clouds, not only block sunlight, it also absorbs radiation both from in-coming solar radiation and from out-going terrestrial radiation. The thusly heated aerosol particles transfer that heat to the surrounding atmosphere which increases its temperature and reduces the adverse temperature gradient relative to air near the surface. The reduction of adverse temperature gradient, as demonstrated by the above classroom-demonstration, concomitantly reduces convective heat transport from the surface. This is the mechanism whereby particulate pollution causes global warming.

Evidence of Global Warming: Two measurements, the daily high temperature and the nightly low temperature, when tracked over time over a large geographic area, provide an independent measure of climate change. The daily high temperature minus nightly low temperature, ($T_{max} - T_{min}$), called the diurnal temperature range (DTR), is essentially independent of any effects of greenhouse gases [396, 427]. Figure 5.13 from Qu et al. [428] presents yearly DTR, T_{max}, and T_{min} mean values over the continental USA throughout most of the 20th century and into the 21st century up to 2010.

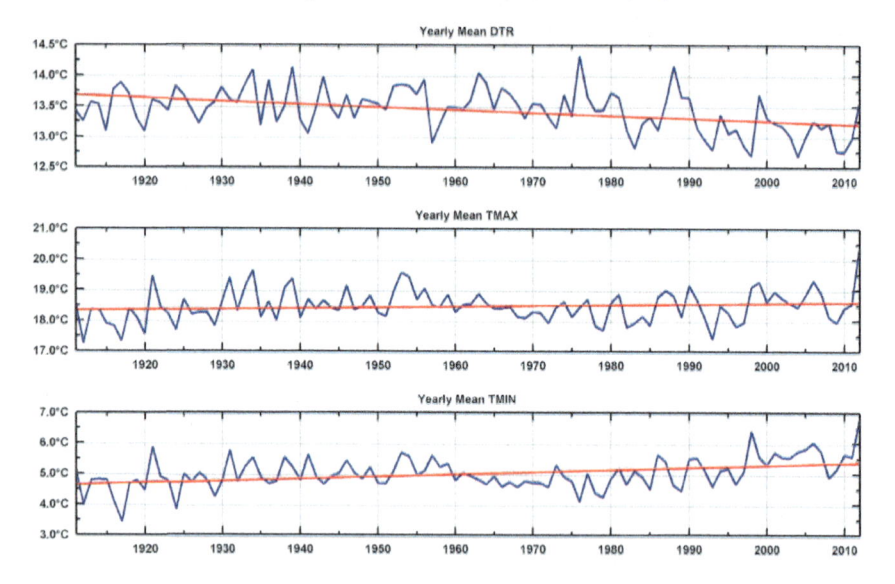

Figure 5.13. Yearly DTR, T_{max}, and T_{min} mean values over the continental USA. The red lines are linear regressions. From [367, 428].

As shown in Figure 5.13, T_{min} increases at a greater rate than T_{max} causing DTM to decrease over time, a phenomena that is observed in many similar investigations [429-432] but not all [433]. Whereas the reduction in T_{max} can be explained by sunlight being absorbed or scattered by particulates or by clouds, the increase in T_{min} is inexplicable within the current consensus-driven climate science paradigm [396].

Further Evidence: The eruption of Mount St. Helens volcano in Washington State (USA) on May 18, 1980 [434] provided an opportunity to assess the short-term influence of tropospheric injection of volcanic particulates [435]. As the volcanic plume passed overhead in the troposphere, daytime temperatures dropped as the sunlight was absorbed and scattered by the particulates; nighttime temperatures, however, increased, and for a few days thereafter remained elevated presumably due to aerosol dust that persisted for a few days before falling to ground [435].

The diurnal temperature range was significantly lessened by the plume, but almost completely recovered within two days [435]. These observations are consistent with (1) the Mount St. Helens aerosol particulates in the plume absorbing long-wave radiation and becoming heated in the atmosphere overhead, (2) the transfer of that heat to the surrounding atmosphere by molecular collisions, (3) the lowering of the atmospheric adverse temperature gradient relative to Earth's surface, (4) the consequent reduction of atmospheric convection, and (5) concomitant reduction of convection-driven surface heat loss, which is evident by the increase in T_{min} [15, 369, 370, 436].

Relying on computer models, the climate science community failed to understand the significant role of atmospheric convection, modulated by aerosol particles, in heat removal from Earth's surface. They seem unaware that convection-efficiency reduction is caused by atmospheric heating via aerosol particulate heating. For example, the explanation proffered for the Mount St. Helens volcanic plume nighttime heating is *"at night the plume suppressed infrared cooling or produced infrared warming"* [435] – which simply does not make sense.

Clouds and tropospheric particulate pollution thus are common factors that affect the diurnal temperature range, DTR. Hypothetically, one might imagine a more-or-less constant DTR if there was no human-caused particulate-pollution. But that is not the case. Instead one observes the consistent decrease in DTR over time, driven by the consistent increase in nighttime T_{min} (Figure 5.13), which, in light of the evidence described above, points to particulate pollution as the principal cause of global warming.

The idea that tropospheric particulates reduce atmospheric

convection received further support by the long-duration series of radiosonde and aethalometer investigations undertaken by Talukdar et al. [437]. Their measurements demonstrated that higher amounts of tropospheric black carbon aerosols can disturb the normal upward movement of moist air by heating up the atmosphere, resulting in a decrease in the atmospheric convection parameters associated with the increase in concentration of black carbon aerosols.

During summer months, Saharan-blown dust, covers an area over the tropical ocean between Africa and the Caribbean about the size of the continental United States [408, 438, 439]. The dust-layer extends to an altitude of 5-6 km; measurements indicate greater dust density and associated haziness at 3 km than at the surface [439]. The warmth of the upper portion of the Saharan-blown dust layer is a consequence of its origin over the Sahara, but the warmth is maintained by the absorption of solar radiation by the dust [438], which is known to contain radiation-absorbing iron oxide [440, 441] that, when incorporated in water-bodies, initiates harmful algae blooms [341, 442-444].

As noted by Prospero and Carlson [439]: " ... *the warmth of the Saharan air has a strong suppressive influence on cumulus convection ....*" Wong and Dessler [445] also recognized the suppression of convection over the tropical North Atlantic by the Saharan-dust air layer. Neither these nor other investigators [438] recognized the global generality, namely, that reduction of convection-efficiency occurs as a consequence of reducing the adverse temperature gradient through aerosol particulate heating [15, 368-370].

Convection occurs *throughout the troposphere*, with differing degrees of scale, both geographically and altitudinally, and with various modifications caused by atmospheric circulation and lateral flow. Convection-efficiency in all instances, however, is a function of the prevailing adverse temperature gradient. Aerosolized particles, heated by solar radiation and/or terrestrial radiation, rapidly transfer that heat to the surrounding atmosphere, which reduces the adverse temperature gradient relative to the surface and, concomitantly, reduces surface heat loss and thereby over time causes increased surface warming [370]. The same particulate-pollution-driven process operates locally, as in the case of urban heat islands [409, 446-449], regionally, and globally.

My new climate science paradigm: *Particulate pollution, not anthropogenic carbon dioxide, is the principal cause of global warming* [15, 368-370].

For nearly four decades, the United Nations through its Intergovernmental Panel on Climate Change has been extracting vast sums of money from wealthy nations to promote the idea that climate change – global warming – is caused by anthropogenic fossil-fuel-produced carbon dioxide. The goal seems to be to control or shut down fossil fuel use in favor of renewable energy sources to ameliorate global warming. Yet through its ENMOD treaty the United Nations has simultaneously sanctioned secret activities that in fact *cause* global warming [15, 23, 318, 368-371]. This must be the greatest scientific fraud ever perpetrated.

A "prospectus," in a medical or public health journal, is a short article that provides early warning of a potential health crisis. We submitted just such a prospectus to the United Nations' *Bulletin of the World Health Organization* warning of the potential health crisis posed by aerosolized coal fly ash being jet-sprayed into the lower atmosphere. That prospectus was rejected *without* review by the *Bulletin of the World Health Organization* [450]. This unwarranted rejection without peer-review occurred not once but twice.

The immediate rejections of this health crisis early warning by the *Bulletin of the World Health Organization* [450] clearly points to United Nations involvement in said global environmental modification activity. There are other bad actors as well: The Lancet and the New England Journal of Medicine likewise rejected our public health warning without peer-review.

In 1968, MacDonald [360] accurately forecasted the military's decision to weaponize the environment for national purposes. He failed, however, to realize that sovereign nations and their militaries, mainstream media, tech giants, super-rich elites, and complicit government agencies, could and would be *co-opted* by a secret international agreement to wage *de facto* war on planet Earth and its biogeochemical processes. In order to undermine those individual states' authority to protect their own biota and natural environment, ENMOD coerces all signatory nations, 78 of them, them to participate in non-disclosed, wholesale global modification.

To reiterate: Solid and/or liquid particles, typically ≤ 10 μm across, in the troposphere originate from a variety of sources including, for example, moisture condensation [451], incomplete biomass burning, combustion of fossil fuels, volcanic eruptions, wind-blown road debris, sand, sea salt, biogenic material [452], pyrogenic coal fly ash [453-456], and deliberate human aerosol pollution activities [331, 332, 450, 457].

The one generalization that can be made is that virtually all tropospheric aerosol particulates, including cloud droplets and their aerosol components, absorb short- and long-wave solar radiation, and absorb long-wave radiation from Earth's surface and become heated.

Whereas the methodology utilized by the IPCC and climate science community has focused primarily on the problem of sun-Earth radiation balance and departures therefrom, my focus has been on understanding the processes involved in the disposition of absorbed heat, notably the consequences of particulate pollution on atmospheric convection, which is a primary mechanism for maintaining Earth's habitable surface temperature [15, 367-370].

Life on Earth is possible because of the natural balance of interactions by and between myriad biota and the physical processes of their environments. For more than three billion years, as long as life has existed on Earth, this balance has maintained itself, despite natural environmental variability, without human intervention. Humans can learn to adapt to natural variability and can prosper. But *unnatural* variability is not only unhealthy, but potentially devastating to humans and other biota.

Anthropogenic (human-created) global warming is one particularly devastating form of *unnatural* variability. Its cause, however, has been grossly misunderstood. It is neither necessary to trap and sequester carbon dioxide [458] nor to emplace particles in the upper atmosphere (stratosphere) to reflect sunlight [459], a disastrous "cure" that in the extreme might even initiate a new ice age. Significantly reducing tropospheric particulate pollution, as real scientific evidence indicates, will quickly lead to reduced global anthropogenic warming on a time scale of days to weeks. Moreover, reducing tropospheric particulate pollution will lead to great improvements in human and environmental health.

Air pollution is the leading environmental cause of disease and death worldwide, and it is increasing at an alarming rate [460]. Exposure to air pollution particles is a significant risk factor for premature death, including ischemic heart disease, chronic obstructive pulmonary disease, and respiratory infections [461]. Long-term, cumulative exposure to fine particulate matter in the United States is associated with all-cause mortality, cardiovascular disease, and lung cancer [462]. In recent years, emerging evidence from clinical, observational, epidemiological and experimental studies strongly suggest that Alzheimer's Dementia, Parkinson's, and thrombotic stroke are associated with ambient air pollution [463]. Children residing in highly polluted urban environments were found to have cognitive deficits, and the majority of them showed brain abnormalities on MRI [464]

The combined-causes of air pollution and runaway global warming are modifiable in a short time-frame by reducing particulate pollution. But corrective actions hinge on embracing the real climate-science paradigm-shift [15, 367-370] and not continuing to promote flawed and harmful political dictums that serve vested self-interests. In the light of real scientific truth, government officials should strive for international cooperation at all levels of authority.

New Stratospheric Ozone Depletion Paradigm: Earth's biosphere is collapsing at an unprecedented rate, including and especially the stratospheric ozone layer that shields surface life from the deadly ultraviolet solar radiation. That collapse, which has been progressing for decades, is due to both deliberate and unintentional human activity. Discovering the causes of biosphere collapse, we submit, should be the highest priority for scientists. But all too often, scientists continue to plod along unquestioningly working in problematic paradigms, while ignoring paradigm shifting discoveries [235, 236].

In the following we review compelling published evidence that supports our contention that aerosolized coal fly ash particles are the main agents responsible for stratospheric ozone depletion, not chlorofluorocarbon gases [337-339, 465, 466].

Aerosolized coal fly ash particles, uplifted to the stratosphere, not only serve as ice-nucleating agents, but are trapped and concentrated in stratospheric clouds, including Polar Stratospheric Clouds. In springtime, as stratospheric clouds begin to melt/evaporate, said ozone-consuming coal fly ash particles are released making them available to react with and consume stratospheric ozone. Ceasing to contaminate the environment with aerosolized coal fly ash will decrease stratospheric ozone destruction, will reduce global warming, and will significantly improve human and environmental health.

Background Mistakes: Since about 1913, there was much speculation as to why the intensity of solar radiation measured at Earth's surface nearly vanished at wavelengths less than at about 290 nm. In 1925, Fabry [467] determined that laboratory-measured light absorption by ozone (O_3) matched the solar spectral decrease observed.

In 1985, Farman et al. [468] reported that total ozone levels over Antarctica during the month of October had steadily decreased since 1970. In 1986, from satellite measurements Stolarski et al. [469] showed that the "ozone hole" covers all of Antarctica and corresponds to the region enclosed by the southern polar vortex [470]. The cause of the Antarctic "ozone hole" was a great mystery [471].

Considerable efforts were expended to determine the cause of stratospheric ozone depletion. The proposed chemical species typically involved gas-phase reactions, usually involving halogens and photodissociation [472, 473].

In 1974, Rowland and Molina [472], advanced the theory that gaseous substances containing chlorine and fluorine (so-called CFCs) were destroying the stratospheric ozone layer. They reported that these molecules would not break down in the atmosphere and eventually find their way to the stratosphere where they would be photolyzed to release reactive chlorine, which depletes ozone. Ozone depletion by CFCs presumably would occur under sunlit conditions in the upper stratosphere (30-50 km), not in the lower stratosphere, where most of the ozone resides. For their work, Roland and Molina were awarded the 1995 Nobel Prize.

The overall ozone depletion was expected to be 5-10%, not enough to explain the newly discovered Antarctic ozone hole. Homogeneous (gas phase) chemistry could not account for the ozone loss [474]. In 1986, Solomon and coworkers argued that newly discovered Polar Stratospheric Clouds (PSCs) in the extremely cold polar lower stratosphere provide reaction sites for heterogeneous chemical reactions between the relative inert chlorine gases HCl and $ClONO_2$. She hypothesized that Antarctic Polar Stratospheric Clouds were electrostatically attracting CFCs and providing them sites in the form of ice crystals, on which the Rowland/Molina proposed ozone-depleting reactions could rapidly take place [475].

In 1989, the United Nations (UN) formally adopted the *"Montreal Protocol on Substances that Deplete the Ozone Layer"* that included regulating numerous halogen-containing chemicals that readily form gases [476]. In doing so, the United Nations *decreed* that said halogen-containing chemicals were in fact the main cause of ozone depletion. The United Nations by international agreement phased out the production of CFCs to remedy the problem of stratospheric ozone production. However, the United Nations was incorrect.

The rapidly increasing penetration of ultraviolet B and C radiation to

Earth's surface portends a potentially dire depletion of stratospheric ozone [340, 477, 478]. The clearly visible destructive effects of UV on global ecosystems including forests and coral reefs should be a warning sign that stratospheric ozone depletion may be the biosphere's most imminent threat [340] (Figure 5.14).

Figure 5.14. Comparison of sunny and shaded sides of two trees. Left, Torrey Pine; Right, Gumbo-Limbo. From [349].

Recently, not only has a large ozone hole been observed in the Arctic [479], but in the tropics as well [480]. From these indications, as well as from the data shown in [481-483], one thing is abundantly clear: The Montreal Protocol misdiagnosed the cause of stratospheric ozone depletion, and its sanctions on chlorofluorocarbons (CFCs) have not been the solution (Figure 5.15).

Seasonal Changes in the UV Index

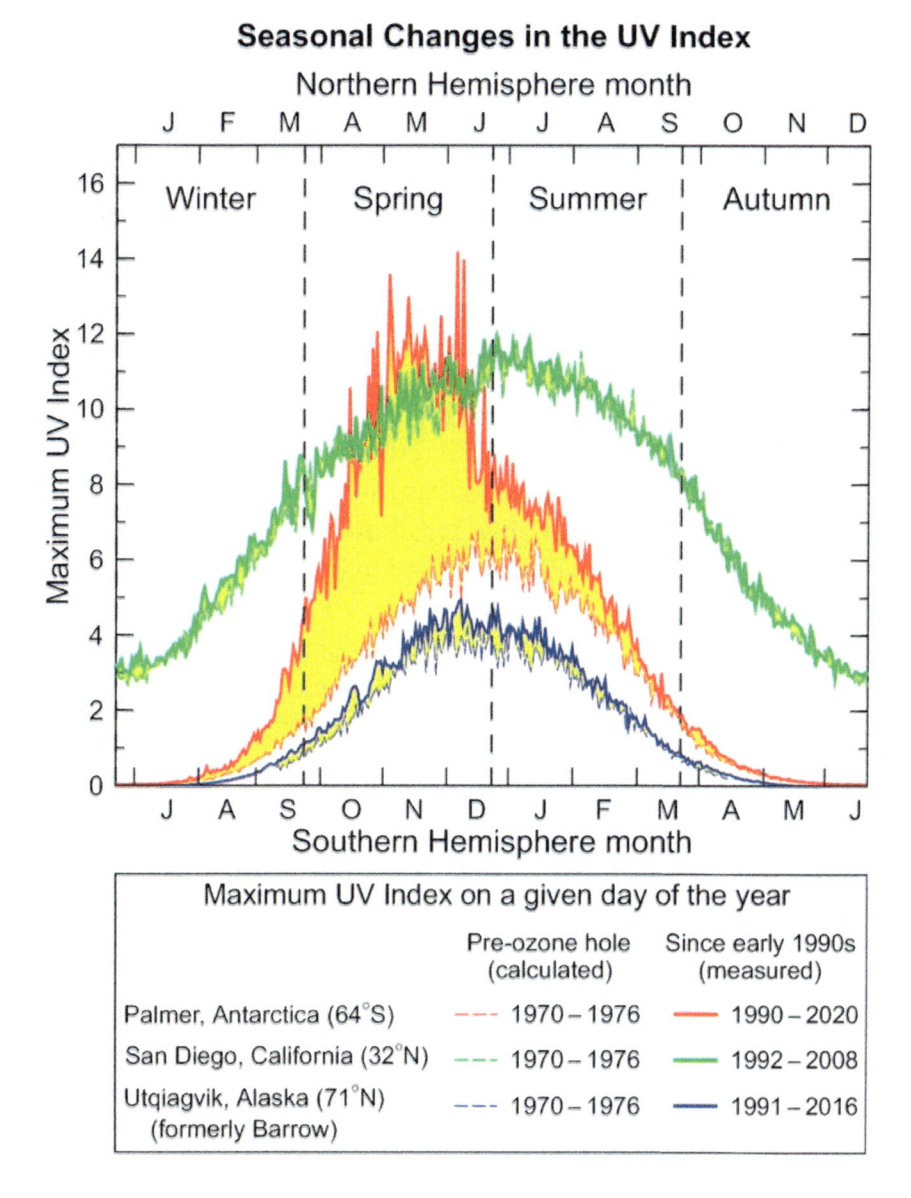

Figure 5.15. The current and historical status of stratospheric ozone revealed by measurements of UV Index disclosing the worsening of the Antarctic ozone hole. From [481].

Recent discoveries about the causes of Earth's Great Extinctions suggest another more likely cause of stratospheric ozone depletion, coal fly ash. Earth's great extinctions correlate with epic volcanic phenomena called Large Igneous Provinces (LIPs) [484]. The Permian Extinction (The Great Dying) 250 million years ago coincided with the Siberian Traps LIP, a massive outpouring of lava and intrusion of underground magma which mixed with thick coal seams. This hot coal-basalt mixture extruded at numerous surface locations, producing multiple plumes of pyroclastic fly ash, soot, sulfate and basaltic dust which ascended to the upper atmosphere [485]. This material was dispersed globally and the resulting char deposits in Permian-aged rocks were found to be remarkably similar to modern coal fly ash [486].

The Permian Extinction was characterized by high levels of carbon dioxide, methane, and rapid global warming to levels lethal to most living organisms [487]. The idea was put forward that a period of deadly ultraviolet radiation stress may have resulted from stratospheric ozone depletion due to this outpouring of hydrothermal organo-halogens [488], but perhaps better described as from natural formations of aerosolized coal fly ash. The Cretaceous-Tertiary (K-T) extinction 65 million years ago is known for the disappearance of dinosaurs and the Chicxulub (asteroid) impact. However, recent scientific evidence has linked this mass extinction to resurgent Deccan Trap LIP volcanism [489].

We are already well into the first anthropogenic extinction event in which coal fly ash is an integral, crucial part, not the least of which, we submit, is the destruction of stratospheric ozone which shields surface life from harmful solar ultraviolet radiation.

During industrial coal burning, the heavy ash settles beneath the burner; the light ash, coal fly ash, forms in the gases above the burner and exits the smokestacks, unless, as in Western nations, it is trapped by electrostatic precipitators and sequestered. Even so, ultrafine aerosols from coal burning are likely to escape electrostatic precipitators [490] or be wind-blown from sequestration areas [491]. But the most devastating adverse consequence for life on this planet is the deliberate, covert, near-daily, near-global jet-emplacement of particulates, evidenced as coal fly ash, into the upper troposphere [331, 332, 336, 362].

Coal Fly Ash in the Stratospheric Polar Vortex: Coal fly ash effectively nucleates ice at conditions relevant to mixed phase clouds. Enhanced ice nucleation by coal fly ash aerosol particles is initiated by their porous structures [492]. In the scientific literature, coal fly ash particles are often classified or confused with mineral dust particles. The majority of cirrus clouds freeze, or nucleate around two types of seeds, "mineral dust" and metallic aerosols, presumably with important contributions from coal fly ash to both categories [493]. For example, particles captured from tropospheric ice nuclei consisting of nanometer carbon balls [494] are similar to carbon balls extracted from coal fly ash [495, 496].

Carbon nanoparticles from coal fly ash occur in a variety of forms [497-511] and nanoparticle compositional ranges [512-519] some of which have been observed in the polar stratosphere [520]. Nanoparticles, lofted into the stratosphere [521-523], display a range of compositions characteristic of coal fly ash, as illustrated by particles captured from Polar Stratospheric Clouds within the Arctic vortex [524].

Coal Fly Ash Kills Ozone: Ozone is destroyed by reaction with halogens [525, 526]. Coal fly ash contains the halogen elements, bromine, chlorine, fluorine, and iodine [527]. Coal burning in China led to an unexpectedly large atmospheric component of reactive bromine and chlorine in the atmosphere [528].

Experiments are sometimes made to render coal fly ash safer and more amenable for commercial use, for example, as a component of cement. Experiments that employ ozone provide important information as to the ability of coal fly ash to destroy ozone. For example, the surfaces of coal fly ash carbon particles are oxidized by ozone [529] demonstrating that coal fly ash carbon particles kill ozone. Similar investigations also indicate that other substances in coal fly ash kill ozone [530, 531].

Inferences regarding ozone destruction by the components of coal fly ash can be made on the basis of ozone destruction by similar compounds: Ozone is consumed by reaction with carbon [532, 533]. Ozone is also consumed by reactions with mineral oxides [534-538].

Furthermore, ozone is consumed by reactions with oxides of iron and manganese [539, 540]. Additionally, ozone is consumed by reactions with metals [541, 542] and noble metals [540, 543]. All of these substances occur in coal fly ash nanoparticles.

Our published evidence, cited above, provides compelling evidence that coal fly ash particles, not chlorofluorocarbons (CFCs), are the primary cause of stratospheric ozone depletion, through numerous ozone-killing reactions, graphically illustrated in Figure 5.16 from [337].

Coal Fly Ash Ozone Killers
Carbon: Char, Soot, Nanotubes, Nanoballs, etc.
Halogens: Chlorine, Bromine, Fluorine, Iodine
Iron Oxides, Manganese Oxides, Mineral Oxides
Metals, Noble Metals, and Mixed Metals
Many Other Possible Ozone Killers in Coal Fly Ash

Stratospheric Cloud

Coal Fly Ash Chemtrails

Coal Fly Ash Dump   Coal Fly Ash Exhaust

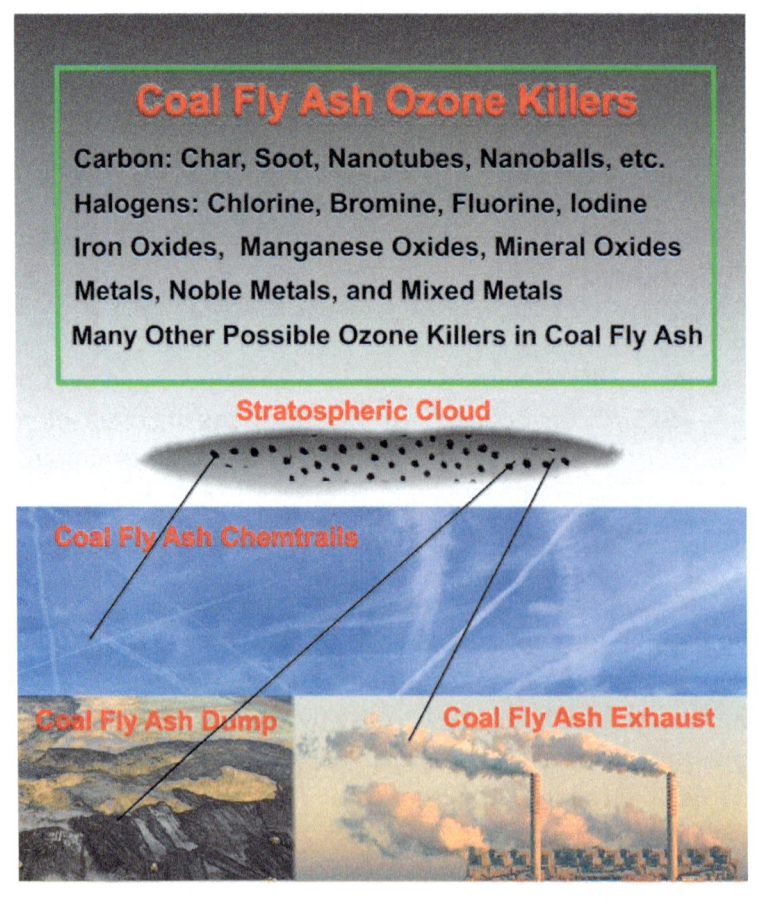

Figure 5.16. Graphic illustrating the major sources of aerosolized coal fly ash lofted into a particle laden polar stratospheric cloud, and some of the many components of coal fly ash that directly kill ozone. From [337].

Polar Stratospheric Clouds and Ozone Depletion: On the basis of three consecutive years of observations, Hamill et al. [470] notes: "[W]e *show that the evaporation of the* [Antarctic Polar Stratospheric] *cloud is highly correlated in time with the decrease in ozone concentration.*" There is general acknowledgement that aerosol particles can serve as cloud nuclei, however, nucleation is typically the extent of discussion. Our experience with aerosolized coal fly ash particles brought to ground by snowfall sheds some light on the connections between Polar Stratospheric Clouds and stratospheric ozone depletion.

For decades, with increasing frequency and geographic range, particulate matter has been jet-sprayed into the troposphere (Figure 5.1). Internationally, officials decline to provide either the composition or the intent of the tropospheric particulate emplacement, and falsely assert that the jet-trails are harmless ice-crystal contrails [321, 332].

Mark Whiteside and I published evidence that coal fly ash is the main aerosolized particulate jet-sprayed into the troposphere [331, 544, 545] by comparing element ratios relative to barium in rainwater and melted snow with corresponding ratios measured in the lixiviate of coal fly ash leaching experiments [327, 334]. We further presented evidence [336] that tropospheric post-chemtrail snowfalls can collect and bring down coal fly ash aerosol particulates in a manner similar to the physical-chemical technique called co-precipitation [546].

In northern regions in the springtime, as the snow begins to melt, it releases the trapped coal fly ash particles which descend and are re-trapped on the underlying snow mold. These observations suggest a commonality in behavior that is applicable to Polar Stratospheric Clouds and ozone destruction.

Coal fly ash particles, lofted into the stratosphere, not only serve as ice-nucleating agents, but are further trapped by clouds, including Polar Stratospheric Clouds. In springtime, the icy stratospheric clouds melt/evaporate releasing their trapped coal fly ash particles, and making those ozone-consuming coal fly ash particles readily available for reaction with and destruction of ambient stratospheric ozone.

We have presented compelling evidence that supports our contention that aerosolized coal fly ash particles are the main agents responsible for stratospheric ozone depletion, not chlorofluorocarbon gases [337-339]. Aerosolized coal fly ash particles, uplifted to the stratosphere, not only serve as ice-nucleating agents, but are trapped and concentrated in stratospheric clouds, including Polar Stratospheric Clouds. In springtime, as stratospheric clouds begin to melt/evaporate, said ozone-consuming coal fly ash particles are released making them available to react with and consume stratospheric ozone.

The Big Ugly Picture: Aerosolized coal fly ash particles are responsible, not only for the destruction of stratospheric ozone, which shields surface-life from deadly solar ultraviolet radiation, but for harm to human and environmental health. Published scientific and medical articles implicate aerosolized coal fly ash in neurodegenerative disease [343], COPD and respiratory disease [345, 450], lung cancer [346], cardiovascular disease [342], COVID-19 and immunopathology [348, 547].

Aerosolized coal fly ash contributes to global warming [367], disrupts habitats [545], contaminates the environment with mercury [336], decimates populations of insects [352], bats [353], and birds [335]. Aerosolized coal fly ash also kills trees [349, 350], exacerbates wildfires [320], enables harmful algae in our waters [341], and, as described here and in the scientific literature [337-339, 465, 466], destroys the stratospheric ozone layer that shields surface-life from the sun's deadly ultraviolet radiation. Despite the official narratives of "ozone recovery" due to the Montreal Protocol, stratospheric ozone levels continue to decline [548]. Ozone depletion has already led to an alarming increase in deadly ultraviolet radiation B and C penetration to Earth's surface, with increasingly apparent devastation to both plants and animals [340].

To the best of our knowledge, all of our scientific discoveries along these lines have been ignored by the Deep State American and European scientific communities in stark contradiction to long-established scientific and ethical principles.

An Even Uglier Picture: In 1968 MacDonald [355] foresaw the possibility that in the future the world's militaries might develop the means to trigger on-demand covert environmental modifications to cause storms, floods, droughts, earthquakes, and tidal waves. As he explained: *"The key to geophysical warfare is the identification of the environmental instabilities to which the addition of a small amount of energy would release vastly greater amounts of energy."*

As reviewed [358], MacDonald [355] discussed purposefully triggering instabilities in such large-scale natural systems as the weather, the climate, the oceans, and the human brain, including such phenomena as hurricanes, earthquakes, and tsunamis for use in warfare. He was aware, considering the limitations of geophysical understanding, that one should also anticipate unforeseen adverse consequences that could arise from deliberately disturbing complex natural systems that have unknown "tipping points."

In 1997, U.S. Secretary of Defense William Cohen confirmed military involvement in just the activities Macdonald had predicted as evidenced by Cohen's direct statement [549]: *"Others are engaging...in an eco-type terrorism whereby they can alter the climate, set off earthquakes, [and] volcanoes remotely through the use of electromagnetic waves....It's real, and that's the reason why we have to intensify our efforts."*

An unclassified final document of findings and determinations signed by 90 deputies, released by the Russian State Duma in 2002, and sent to President Putin, the United Nations, and others warned that Americans were experimenting with a new type of geophysical weapon without international control, referred to as HAARP, the acronym for *High Frequency Active Auroral Research Program* [550]. The signatories demanded that an international ban be placed on such large-scale geophysical experiments, and expressed concern for the development of the ionosphere heater weapon in Alaska (USA), Norway, and Greenland noting [550]: *"The significance of this qualitative leap could be compared to the transition from cold steel to fire arms, or from conventional weapons to nuclear weapons. This new type of weapons differs from previous types in that the near-Earth medium becomes at once an object of direct influence and its component."*

Method and Means for Triggering or Inducing Earthquakes: Rocks under stress beginning to fracture generate electromagnetic radiation [551-554]. So it is also with earthquakes [555-557].

Earthquakes result from the catastrophic release of stress that had developed as tectonic plates grated together. Numerous studies, e.g. [552, 555-561], have documented that, as earthquakes are about to occur, electromagnetic radiation is produced at *ultra-low frequencies* (ULF, <3 Hz), *extremely low frequencies* (ELF, 3-3000 Hz), and *very low frequencies* (VLF, 3-30 kHz). As earthquakes are about to occur, electromagnetic radiation is observed at or above the surface [552, 555-561]. This observation demonstrates that electromagnetic radiation of said frequencies can travel through the ground to the depth at which some earthquakes occur. The specific mechanism is yet unknown, but presumably involves micro-fractures forming a resonant cavity or antenna.

A radiation producing resonant device will absorb energy from an external energy source at the same resonant frequency. A vibrating tuning fork will initiate vibrations in a nearby identical tuning fork. Thumping or circling the rim of a crystal glass with a moist finger will cause a crystal glass to resonate in the audio frequency range; a singer's voice at its resonant frequency will cause a crystal glass likewise to resonate and even to shatter [562]. A radio frequency circuit tuned to the frequency of external electromagnetic radiation will absorb energy from external electromagnetic radiation of the same frequency [563]. The same general principle applies to the electromagnetic radiation producing resonance of highly-stressed fault lines in the Earth's crust.

Without ambiguity it follows that ULF/ELF/VLF electromagnetic radiation of the appropriate range of frequencies, applied to a highly stressed, unstable junction of tectonic plates, can trigger or induce an earthquake.

The ULF/ELF/VLF electromagnetic radiation produced in the early stages of earthquake formation has been observed by satellites [552, 555-561], evidence that said radiation propagates in the Earth-ionosphere waveguide [564]. Significantly, **ULF/ELF/VLF electromagnetic radiation** produced by modulated high-frequency electromagnetic heating of the ionosphere [565-568] with HAARP or similar ionosphere heaters, such as **Tromsø in Norway [569]**, can be steered [570, 571] and **can travel thousands of kilometers in the Earth-ionosphere waveguide [572-575] to reach a point of geophysical seismic instability to trigger or induce an earthquake [576-578].**

Information about the operation of ionosphere heaters and their capabilities to induce or trigger earthquakes is not only withheld from the public, but the public is deceived by Deep State lies.

Mick West is well-known for disseminating false information related to Deep State covert activities [579, 580], such as the covert widespread jet-spraying of toxic particulates into the air we breathe, aka chemtrails [310, 581, 582], destruction of the World Trade Center towers [583-585], and activities of the Central Intelligence Agency [586, 587]. Mick West reportedly stated that HAARP "can only affect a small amount of the air above it" [588]. West's illustration from [589] is shown in Figure 5.17a. Contrast scientific fact to West's Figure 5.17a deceptive caption, *"This is roughly the extent of HAARP effect. It heats the ionosphere about 70km above the ground directly above its Alaska location"*. Scientists acknowledge and illustrate in peer-reviewed scientific literature that electromagnetic waves produced by HAARP and other ionosphere heaters *"can travel thousands of kilometers"* [572-575], as illustrated in Figure 5.17b from [575] and in Figure 5.17c from [574].

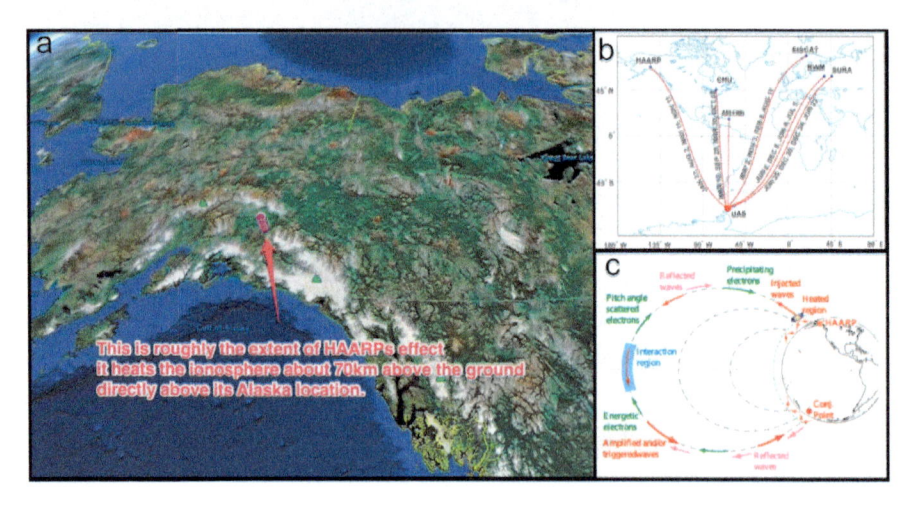

Figure 5.17. a) Mick West's disinformation graphic contrasted to peer-reviewed, scientific graphics b) and c) showing that electromagnetic waves produced by HAARP and other ionosphere heaters can travel *"thousands of kilometers"*.

Figure 5.18 (left) shows the HAARP ionosphere heater in Gakona, Alaska (USA). Its 180 units are capable of producing a total radiated power level of 3.6 MW with an effected radiated power of approximately 575 MW [574, 590]. Figure 5.18 (right) shows one of the other ionosphere heaters, this one in Tromsø, Norway, with an effective radiated power of 300 MW [567, 569, 591] which, like HAARP, is capable of triggering or inducing earthquakes [576-578].

Figure 5.18. Left: High Frequency Active Auroral Research Program (HAARP) phased-array facility in Gakona, Alaska (USA). Right: Norwegian Tromsø phased-array ionosphere heating facility.

The individual antennas, shown in Figure 5.18, operate at different phases designed to thus yielding a high-powered, narrowly focused directional output. These two dimensional phased-arrays are now as obsolete as dinosaurs. A much larger, more powerful, and far more directional, three dimensional phased-array became operational in December 2010 in Antarctica [592, 593].

Project IceCube, a massive neutrino observatory constructed in the ice in Antarctica by the US National Science Foundation, became operational in December 2010 [594] (Figure 5.19).

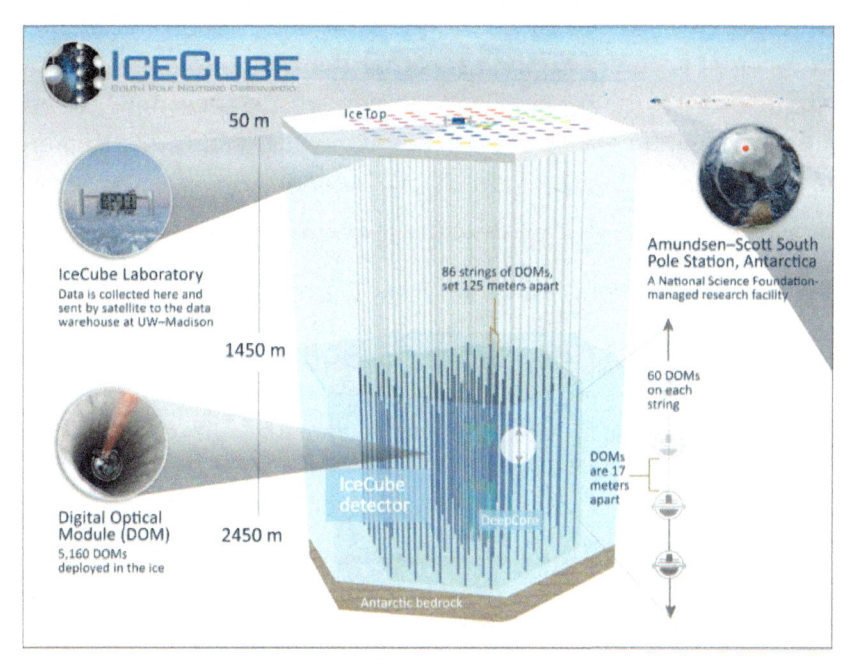

Figure 5.19. Schematic representation of the passive portion of Project IceCube neutrino detector [594]. Not shown is the active portion, a high-powered, three-dimensional phased-array transmitter [592, 593].

In 2023, Eric Hecker, a contractor-employee at the IceCube facility during the start-up period in early 2011, in sworn testimony before the US Senate Intelligence Committee, disclosed the other secret and diabolical feature of Project Ice Cube [592, 593]. In addition to the being a passive neutrino detector, the system is wired to be an active high-powered, three-dimensional phased-array transmitter, described as being *"HAARP on steroids."*

Deliberately Triggered New Zealand Earthquake: Evidence indicates that the first deliberate triggering or inducing an earthquake by said Antarctic IceCube active high-powered, three-dimensional phased-array transmitter occurred on February 22, 2011 shortly before 12:30pm (local time) in Christchurch, New Zealand.

In 2011, an email from Jacob J. (Jake) Sullivan (recently National Security Advisor to the Biden Administration) to then Secretary of State Hillary Clinton [595] indicates that the time of a 6.3 magnitude earthquake in New Zealand was known in advance and provides compelling evidence that the earthquake was deliberately triggered or induced (Figure 5.20).

> From: Sullivan, Jacob J SullivanJJ@state.gov
> Sent: Monday, February 21, 2011 7:35 PM
> To: H; Mills, Cheryl D; Abedin, Hume; Reines, Philippe I
> Subject: Fw: 6.3 magnitude earthquake in Christchurch, New Zealand
> And on cue...[emphasis added]

Figure 5.20. Text of email released by Wikileaks demonstrating foreknowledge of the time of Christchurch earthquake thus evidence that it was deliberately triggered or induced [595].

In 2023, further evidence that said Christchurch, New Zealand earthquake was deliberately triggered or induced by said Antarctic IceCube active high-powered, three-dimensional phased-array transmitter was provided in sworn US Senate testimony by Eric Hecker [592, 593].

Said earthquake occurred at a depth of 5.9 km, later estimated to be magnitude 6.1 [596]. There were 185 deaths as a result of the earthquake, and 6,659 major injuries (in the first 24 hours) [597]. A delegation of 9 U.S. Congressmen departed the city before the quake occurred [598].

Diabolical Activity: Nikola Tesla first foresaw the potential of being able to transport energy through the air without wires. The technological feasibility of Tesla's vision has now been realized, not for the benefit of humanity, but for the sociopathic/evil activity of harming humanity and the natural environment. According to Hecker [592, 593], it is a simple process to dial in essentially any location on Earth and the IceCube three-dimensional phased-array transmitter will deliver energy to that spot via radiation propagation in the Earth-ionosphere waveguide. That directed energy then could induce earthquakes, volcano eruptions, or generate hurricanes and storms in ocean waters.

Circumstantial Evidence of Deliberately Induced February 6, 2023 Turkey Earthquake: On February 6, 2023, a powerful 7.8 magnitude earthquake occurred in southern Turkey near the northern border of Syria. This earthquake was followed about 9 hours later by a 7.5 magnitude earthquake approximately 90 km to the north [599]. These earthquakes and aftershocks devastated parts of Turkey and Syria with the death toll reportedly exceeding 55,100, the majority being women and children [600-602], and an estimated 1.5 million people made homeless [603] (Figure 5.21). News reports before and immediately following the event, together with scientific facts and circumstances presented below, warrant a full, thorough, and complete investigation that said February 6, 2023 earthquake was deliberately triggered or induced and constitutes war crimes and crimes against humanity.

Figure 5.21. Scenes of damage from the February 6, 2023 earthquake in Turkey.

A February 3, 2023 Aljazeera report, *three days before the earthquake in Turkey*, stated [604]: *"Earlier this week nine Western* nations either closed down their consulates in Istanbul or issued travel warnings to citizens visiting Turkey, citing security threats....Turkey says Western nations, including the United States and Germany, have not given it any information to back up their assertions that security threats had prompted them to close their missions in the country.... The measures angered Turkey, which on Thursday summoned the countries' ambassadors in protest.... 'They say there is a terror threat ... But when we ask what the source of information was and who the perpetrators of such attacks might be, they did not share any information with our intelligence and security authorities,' Turkish Foreign Minister Mevlüt Çavuşoğlu said ... and 'We have the information that some countries asked others to shut their consulates.'"

As reported February 14, 2023 [605]: "According to Romanian Senator Diana Ivanovici Șoșoacă, [the] earthquake in Turkey and Syria was caused by the U.S."

Speaking before the Romanian Parliament two days after the earthquake [606], the Senator observed:
- 24 hours before the earthquake, 10 Western countries recalled their ambassadors from Ankara.
- 5 days before the earthquake, several Western countries, including Romania, issued a travel warning to Turkey without providing a motivation.
- One minute before seismographers detected the earthquake, Turkish gas and oil pipelines were shut down. (this is the only assertion unverified by [605])

Further evidence that at least the first of the February 6, 2023 earthquakes were deliberately triggered is the speed with which disinformation sources swung into action to deceive the public by providing false and/or misleading opinions, totally devoid of scientific basis, frequently targeting news sources as "conspiracy theorists" [607-611].

Motives, opportunity and predisposition: Some potential motives for punishing Turkey by using the earthquake-triggering geophysical weapon include the following as described [612, 613] with

references [614]:
- Turkey is blocking Sweden's and Finland's membership in NATO.
- Today Turkey supports Russia in the same way as China, supplying Russia with spare parts, including re-exporting equipment from the U.S. to Russian firms targeted by coercive measures unilaterally imposed by Washington.
- Turkey dismissed Brian Nelson, U.S. Treasury Under-Secretary for terrorism and financial intelligence, who demanded an end to Turkey's exports to Russia and to Turkey's opposition to Sweden's accession to NATO.

Jacob J. (Jake) Sullivan, recently National Security Advisor to the Biden Administration, allegedly participated in the allegedly triggered 2011 earthquake in Christchurch, New Zealand as documented in his email to Hillary Clinton [595] and therefore was fully cognizant of the means, and potential method for triggering the February 6, 2023 Turkey earthquake [614]. Moreover, according to Seymour Hersh [615], Jake Sullivan was directly involved in planning the eco-terrorism attack on the Nord Stream 1 and 2 pipelines allegedly perpetrated by the United States and Norway. In that capacity, I allege, Jake Sullivan had the position, predisposition, and opportunity to plan, with other high U.S. officials, an operation to deliberately trigger or induce the February 6, 2023 Turkey earthquake by using ionosphere heaters located in Antarctica and/or elsewhere.

War Crimes and Crimes against Humanity: Common sense and an understanding of human nature would lead one to the conclusion that other earthquakes have been deliberately triggered or induced by said ionosphere heaters. About three months after the Antarctic IceCube three-dimensional phased-array transmitter was completed and the Christchurch earthquake test was deemed a success, a 9.0 magnitude Tohoku earthquake generated a tsunami that destroyed the Daiichi Nuclear power Plant in Fukushima, Japan. Was this earthquake natural or induced? What about other subsequent eatthquakes?

What is the extent of human and environmental suffering caused by covert use of ionosphere heaters and other means and devices to alter Earth's natural processes? The persons responsible for authorizing these activities and those involved in deceiving the public, I allege, should be adjudicated for War Crimes and Crimes against Humanity.

Technology Bill of Rights: Unless global populations, politicians, news media, scientists, and others in our society face the truth of what is happening before their very eyes and collectively demand an end to the technologically-based assault on our environment, replete with its dissemination of false information [359], we will inevitably continue to charge forward in the first ever anthropogenic species extinction.

People throughout the world are in dire need of means to protect themselves and their families – and indeed, all life on Earth – from ruthless transnational technological assault. The wellbeing and very existence of humanity is at stake. In the United States our personal freedoms and properties are protected from government excesses by the Bill of Rights, but Americans and other members of the world community are not protected against multi-dimensional, transnational technological excesses.

In the United States we desperately need a set of new Constitutional Amendments that collectively form a second Bill of Rights, a Technology Bill of Rights, to protect our freedoms, and our right to clean air, water, soil/agriculture/food, and to protect the planetary environment from deliberate perversion and alteration via large-scale environmental modification or geoengineering. Such a new Technology Bill of Rights could serve as a model for other sovereign nations [359]. Nothing less than a cultural awakening and a new, morally grounded humanitarian, scientific revolution will be needed to bring this system of technology protections to fruition.

Without it we are doomed.

REFERENCES AND AUTHORITIES

1. Patel, K.P., *Government Gangsters: The Deep State, the Truth, and the Battle for Our Democracy*. 2023: Simon and Schuster.

2. https://archive.org/stream/cia-director-william-casey-disinformation-program-quote-soruce/CIA%20Director%20William%20Casey%20Disinformation%20Program%20Quote%20Soruce_djvu.txt

3. Suess, E., *Das Antlitz der Erde*. 1885, Prague and Vienna: F. Tempky.

4. Herndon, J.M., *The nickel silicide inner core of the Earth*. Proc. R. Soc. Lond, 1979. **A368**: p. 495-500.

5. Herndon, J.M., *Sub-structure of the inner core of the earth*. Proc. Nat. Acad. Sci. USA, 1996. **93**: p. 646-648.

6. Herndon, J.M., *Nuclear georeactor origin of oceanic basalt $^{3}He/^{4}He$, evidence, and implications*. Proc. Nat. Acad. Sci. USA, 2003. **100**(6): p. 3047-3050.

7. Herndon, J.M., *Whole-Earth decompression dynamics*. Curr. Sci., 2005. **89**(10): p. 1937-1941.

8. Herndon, J.M., *Nuclear georeactor generation of the earth's geomagnetic field*. Curr. Sci., 2007. **93**(11): p. 1485-1487.

9. Herndon, J.M., *Nature of planetary matter and magnetic field generation in the solar system.* Curr. Sci., 2009. **96**(8): p. 1033-1039.

10. Herndon, J.M., *Geodynamic Basis of Heat Transport in the Earth.* Curr. Sci., 2011. **101**(11): p. 1440-1450.

11. Herndon, J.M., *Hydrogen geysers: Explanation for observed evidence of geologically recent volatile-related activity on Mercury's surface.* Curr. Sci., 2012. **103**(4): p. 361-361.

12. Herndon, J.M., *New indivisible planetary science paradigm.* Curr. Sci., 2013. **105**(4): p. 450-460.

13. Herndon, J.M., *Terracentric nuclear fission georeactor: background, basis, feasibility, structure, evidence and geophysical implications.* Curr. Sci., 2014. **106**(4): p. 528-541.

14. Herndon, J.M., *New Concept for the Origin of Fjords and Submarine Canyons: Consequence of Whole-Earth Decompression Dynamics.* Journal of Geography, Environment and Earth Science International, 2016. **7**(4): p. 1-10.

15. Herndon, J.M., *Air pollution, not greenhouse gases: The principal cause of global warming.* J. Geog. Environ. Earth Sci. Intn., 2018. **17**(2): p. 1-8.

16. Herndon, J.M., *Cataclysmic geomagnetic field collapse: Global security concerns.* Journal of Geography, Environment and Earth Science International, 2020. **24**(4): p. 61-79.

17. Herndon, J.M., *Nature of the Universe: astrophysical paradigm shifts.* Advances in Social Sciences Research Journal, 2021. **8**(1): p. 631-645.

18. Herndon, J.M., *Nuclear fission reactors as energy sources for the giant outer planets.* Naturwissenschaften, 1992. **79**: p. 7-14.

19. Herndon, J.M., *Feasibility of a nuclear fission reactor at the center of the Earth as the energy source for the geomagnetic field.* J. Geomag. Geoelectr., 1993. **45**: p. 423-437.

20. Herndon, J.M., *Planetary and protostellar nuclear fission: Implications for planetary change, stellar ignition and dark matter.* Proc. R. Soc. Lond, 1994. **A455**: p. 453-461.

21. Hollenbach, D.F. and J.M. Herndon, *Deep-earth reactor: nuclear fission, helium, and the geomagnetic field.* Proc. Nat. Acad. Sci. USA, 2001. **98**(20): p. 11085-11090.

22. Herndon, J.M., *Causes and consequences of geomagnetic field collapse.* J. Geog. Environ. Earth Sci. Intn., 2020. **24**(9): p. 60-76.

23. Herndon, J.M., *True science for government leaders and educators: The main cause of global warming.* Advances in Social Sciences Research Journal, 2020. **7**(7): p. 106-114.

24. Herndon, J.M., *Humanity imperiled by the geomagnetic field and human corruption.* Advances in Social Sciences Research Journal, 2021. **8**(1): p. 456-478.

25. Herndon, J.M., *Reasons why geomagnetic field generation is physically impossible in Earth's fluid core.* Advances in Social Sciences Research Journal, 2021. **8**(5): p. 84-97.

26. Herndon, J.M., *Whole-Earth decompression dynamics: new Earth formation geoscience paradigm fundamental basis of geology and geophysics.* Advances in Social Sciences Research Journal, 2021. **8**(2): p. 340-365.

27. Herndon, J.M., *New mechanism driving major species extinction events.* European Journal of Applied Sciences, 2024. **12**(1): p. 517-530.

28. Herndon, J.M., *Moon's two faces: near-side/far-side maria disparity.* European Journal of Applied Sciences, 2023. **11**(2): p. 430-440.

29. Herndon, J.M., *Mechanism of solar activity triggering earthquakes and volcanic eruptions.* European Journal of Applied Sciences, 2022. **10**(3): p. 408-417.

30. Herndon, J.M., *Whole-Mars Decompression Dynamics.* European Journal of Applied Sciences, 2022. **10**(3): p. 418-438.

31. Herndon, J.M., *Formation of mountain ranges: Described By Whole-Earth Decompression Dynamics.* Journal of Geography, Environment and Earth Science International, 2022. **26**(3): p. 52-59.

32. Herndon, J.M., *Validation of the protoplanetary theory of solar system formation.* Journal of Geography, Environment and Earth Sciences International, 2022. **26**(2): p. 17-24.

33. Herndon, J.M., *Scientific basis and geophysical consequences of geomagnetic reversals and excursions: A fundamental statement.* Journal of Geography, Environment and Earth Science International, 2021. **25**(3): p. 59-69.

34. Drake, S., ed. *Discoveries and Opinions of Galileo.* 1956, Doubleday: New York. 301.

35. Box, G.E.P., *Empirical Model-Building and Response Surfaces.* 1987: Wiley.

36. Herndon, J.M., *Inseparability of science history and discovery.* Hist. Geo Space Sci., 2010. **1**: p. 25-41.

37. Copernicus, N., *De revolutionibus orbium coelestium.* 1543, Nuremberg: Johannes Petreius. 405.

38. Stimson, D., *The Gradual Acceptance of the Copernican Theory of the Universe*, in *Faculty of Political Science.* 1917, Columbia University: Hanover, New Hampshire. p. 149.

39. Heath, T.L., *The works of Archimedes.* 1897, Cambridge: Cambridge University Press.

40. Yangshen, S., et al., *Plate tectonics of east Qinling mountains, China.* Tectonophysics, 1990. **181**(1-4): p. 25-30.

41. Herndon, J.M., *Fictitious Supercontinent Cycles.* Journal of Geography, Environment and Earth Science International, 2016. **7**(1): p. 1-7.

42. Kuhn, T.S., *The Structure of Scientific Revolutions.* 1962, Chicago, IL, USA: University of Chicago Press.

43. Illy, J., *Einstein's Gyros.* Physics in Perspective, 2019. **21**(4): p. 274-295.

44. Courtillot, V. and J.L. Le Mouël, *The study of Earth's magnetism (1269–1950): A foundation by Peregrinus and subsequent development of geomagnetism and paleomagnetism.* Reviews of Geophysics, 2007. **45**(3).

45. Needham, J., *Science and Civilisation in China, vol. 4, Physics and Physical Technology, Part 1, Physics,.* 1962, New York: Cambridge University Press.

46. Gilbert, W., *De Magnete.* 1600, London: Peter Short. 240.

47. Gauss, J.C.F., *Allgemeine Theorie des Erdmagnetismus: Resultate aus den Beobachtungen des magnetischen Vereins in Jahre 1838.* 1838, Leipzig. 73.

48. Faraday, M., *Experimental researches in electricity, vol. III.* London, UK: Richard Taylor and William Francis, 1855: p. 1846-1852.

49. Oldham, R.D., *The constitution of the interior of the earth as revealed by earthquakes.* Q. T. Geol. Soc. Lond., 1906. **62**: p. 456-476.

50. Elsasser, W.M., *On the origin of the Earth's magnetic field.* Phys. Rev., 1939. **55**: p. 489-498.

51. Elsasser, W.M., *Induction effects in terrestrial magnetism.* Phys. Rev., 1946. **69**: p. 106-116.

52. Elsasser, W.M., *The Earth's interior and geomagnetism.* Revs. Mod. Phys., 1950. **22**: p. 1-35.

53. Kirschvink, J.L., *Magnetite biomineralization and geomagnetic sensitivity in higher animals: an update and recommendations for future study.* Bioelectromagnetics: Journal of the Bioelectromagnetics Society, The Society for Physical Regulation in Biology and Medicine, The European Bioelectromagnetics Association, 1989. **10**(3): p. 239-259.

54. Sequeira, A.M., *Animal Navigation: The Mystery of Open Ocean Orientation.* Current Biology, 2020. **30**(18): p. R1054-R1056.

55. Russell, C., *The solar wind interaction with the Earth's magnetosphere: A tutorial.* IEEE transactions on plasma science, 2000. **28**(6): p. 1818-1830.

56. Marusek, J.A., *Solar storm threat analysis.* 2007: J. Marusek.

57. Oughton, E.J., et al., *A risk assessment framework for the socioeconomic impacts of electricity transmission infrastructure failure due to space weather: An application to the United Kingdom.* Risk Analysis, 2019. **39**(5): p. 1022-1043.

58. Cannon, P., et al., *Extreme space weather: impacts on engineered systems and infrastructure.* 2013: Royal Academy of Engineering.

59. Baldwin, I., *Discovery of electricity and the electromagnetic force: Its importance for environmentalists, educators, physicians, politicians, and citizens.* Advances in Social Sciences Research Journal, 2020. **7**(12): p. 362-383.

60. Williams, T.J., *Cataclysmic Polarity Shift is US National Security Prepared for the Next Geomagnetic Pole Reversal.* 2015, Air Command and Staff Colleage, Maxwell AFB United States.

61. Li, J., et al., *Shock melting curve of iron: A consensus on the temperature at the Earth's inner core boundary.* Geophysical Research Letters, 2020. **47**(15): p. e2020GL087758.

62. Herndon, J.M., *New concept for internal heat production in hot Jupiter exo-planets.* https://arxiv.org/abs/astro-ph/0612603

63. Gutenberg, B., Zeitschrift Geophysik, 1926. **2**: p. 24-29.

64. Mohorovicic, A., Jb. Met. Obs. Zagreb, 1909. **9**: p. 1-63.

65. Lehmann, I., *P'*. Publ. Int. Geod. Geophys. Union, Assoc. Seismol., Ser. A, Trav. Sci., 1936. **14**: p. 87-115.

66. Birch, F., *The transformation of iron at high pressures, and the problem of the earth's magnetism.* Am. J. Sci., 1940. **238**: p. 192-211.

67. Herndon, J.M. and H.E. Suess, *Can enstatite meteorites form from a nebula of solar composition?* Geochim. Cosmochim. Acta, 1976. **40**: p. 395-399.

68. Herndon, J.M., *The chemical composition of the interior shells of the Earth.* Proc. R. Soc. Lond, 1980. **A372**: p. 149-154.

69. Griffin, A.A., P.M. Millman, and I. Halliday, *The fall of the Abee meteorite and its probable orbit.* Journal of the Royal Astronomical Society of Canada, 1992. **86**: p. 5-14.

70. Dawson, K.R., J.A. Maxwell, and D.E. Parsons, *A description of the meteorite which fell near Abee, Alberta, Canada.* Geochim. Cosmochim. Acta, 1960. **21**: p. 127-144.

71. Herndon, J.M. and M.L. Rudee, *Thermal history of the Abee enstatite chondrite.* Earth. Planet. Sci. Lett., 1978. **41**: p. 101-106.

72. Rudee, M.L. and J.M. Herndon, *Thermal history of the Abee enstatite chondrite II, Thermal measurements and heat flow calculations.* Meteoritics, 1981. **16**: p. 139-140.

73. Keil, K., *Mineralogical and chemical relationships among enstatite chondrites.* J. Geophys. Res., 1968. **73**(22): p. 6945-6976.

74. Dziewonski, A.M. and F. Gilbert, *Observations of normal modes from 84 recordings of the Alaskan earthquake of 1964 March 28.* Geophys. Jl. R. Astr. Soc., 1972. **72**: p. 393-446.

75. Dahm, C.G., *New values for dilatational wave-velocities through the Earth.* Eos, Transactions American Geophysical Union, 1934. **15**(1): p. 80-83.

76. Bullen, K.E., *A hypothesis on compressibility at pressures on the order of a million atmospheres.* Nature, 1946. **157**: p. 405.

77. Vidale, J.E. and H.M. Benz, *Seismological mapping of the fine structure near the base of the Earth's mantle.* Nature, 1993. **361**: p. 529-532.

78. Mao, W.L., et al., *Ferromagnesian postperovskite silicates in the D" layer of the Earth.* Proc. Nat. Acad. Sci. USA, 2004. **101**(45): p. 15867-15869.

79. Chandler, B., et al., *Exploring microstructures in lower mantle mineral assemblages with synchrotron x-rays.* Science Advances, 2021. **7**(1): p. eabd3614.

80. Maruyama, S., M. Santosh, and D. Zhao, *Superplume, supercontinent, and post-perovskite: mantle dynamics and anti-plate tectonics on the core–mantle boundary.* Gondwana Research, 2007. **11**(1-2): p. 7-37.

81. Herndon, J.M., *Scientific basis of knowledge on Earth's composition.* Curr. Sci., 2005. **88**(7): p. 1034-1037.

82. Herndon, J.M., *The object at the centre of the Earth.* Naturwissenschaften, 1982. **69**: p. 34-37.

83. Murrell, M.T. and D.S. Burnett, *Actinide microdistributions in the enstatite meteorites.* Geochim. Cosmochim. Acta, 1982. **46**: p. 2453-2460.

84. Galer, S. and R. O'nions, *Residence time of thorium, uranium and lead in the mantle with implications for mantle convection.* Nature, 1985. **316**(6031): p. 778-782.

85. Fermi, E., *Elementary theory of the chain-reacting pile.* Science, Wash., 1947. **105**: p. 27-32.

86. Herndon, J.M., *Examining the overlooked implications of natural nuclear reactors.* Eos, Trans. Am. Geophys. U., 1998. **79**(38): p. 451,456.

87. Rao, K.R., *Nuclear reactor at the core of the Earth! - A solution to the riddles of relative abundances of helium isotopes and geomagnetic field variability.* Curr. Sci., 2002. **82**(2): p. 126-127.

88. Craig, H. and J. Lupton, *Primordial neon, helium, and hydrogen in oceanic basalts.* Earth and Planetary Science Letters, 1976. **31**(3): p. 369-385.

89. Anderson, D.L., *Helium-3 from the mantle - Primordial signal or cosmic dust?* Science, 1993. **261**(5118): p. 170-176.

90. Honda, M. and I. McDougall, *Primordial helium and neon in the Earth—a speculation on early degassing.* Geophysical research letters, 1998. **25**(11): p. 1951-1954.

91. Rasmussen, M., et al., *Olivine chemistry reveals compositional source heterogeneities within a tilted mantle plume beneath Iceland.* Earth and Planetary Science Letters, 2020. **531**: p. 116008.

92. Mundl-Petermeier, A., et al., *Anomalous 182W in high 3He/4He ocean island basalts: Fingerprints of Earth's core?* Geochimica et Cosmochimica Acta, 2020. **271**: p. 194-211.

93. Anderson, D.L., *The statistics of helium isotopes along the global spreading ridge system and the central limit theorem.* Geophys. Res. Lett., 2000. **27**(16): p. 2401-2404.

94. Dodson, A., B.M. Kennedy, and D.J. DePaolo, *Helium and neon isotopes in the Imnaha Basalt, Columbia River Basalt Group: evidence for a Yellowstone plume source.* Earth and Planetary Science Letters, 1997. **150**(3-4): p. 443-451.

95. Honda, M., et al., Geochim. Cosmochim. Acta, 1993. **57**: p. 859-874.

96. Lin, H.-T., et al., *Mantle degassing of primordial helium through submarine ridge flank basaltic basement.* Earth and Planetary Science Letters, 2020. **546**: p. 116386.

97. Jackson, M., J. Konter, and T. Becker, *Primordial helium entrained by the hottest mantle plumes.* Nature, 2017. **542**(7641): p. 340-343.

98. Starkey, N.A., et al., *Helium isotopes in early Iceland plume picrites: Constraints on the composition of high 3He/4He mantle.* Earth and Planetary Science Letters, 2009. **277**(1-2): p. 91-100.

99. Raghavan, R.S., et al., *Measuring the global radioactivity in the Earth by multidectector antineutrino spectroscopy.* Phys. Rev. Lett., 1998. **80**(3): p. 635-638.

100. Raghavan, R.S., *Detecting a nuclear fission reactor at the center of the earth.* https://arxiv.org/abs/hep-ex/0208038

101. Domogatski, G.V., et al., *Neutrino geophysics at Baksan I: Possible detection of georeactor antineutrinos.* Physics of Atomic Nuclei, 2005. **68**(1): p. 69-72.

102. Fogli, G., et al., *KamLAND neutrino spectra in energy and time: Indications for reactor power variations and constraints on the georeactor.* Physics Letters B, 2005. **623**(1-2): p. 80-92.

103. Smirnov, O., *Experimental aspects of geoneutrino detection: Status and perspectives.* Progress in Particle and Nuclear Physics, 2019. **109**: p. 103712.

104. Кузьмичев, Л., *Нейтринная астрофизика. Характеристика нейтринного импульса»*, НИИЯФ МГУ, 2019.

105. Domogatski, G., et al., *Neutrino geophysics at Baksan I: Possible detection of Georeactor Antineutrinos.* https://arxiv.org/abs/hep-ph/0401221v1

106. Araki, T., et al., *Experimental investigation of geologically produced antineutrinos with KamLAND.* Nature, 2005. **436**: p. 499-503.

107. McDonough, W.F., *Earth sciences: Ghosts from within.* Nature, 2005. **436**: p. 467-468.

108. Gando, A., et al., *Reactor on-off antineutrino measurement with KamLAND.* Physical Review D, 2013. **88**(3): p. 033001.

109. Agostini, M., et al., *Comprehensive geoneutrino analysis with Borexino.* Physical Review D, 2020. **101**(1): p. 012009.

110. Herndon, J.M., *Maverick's Earth and Universe.* 2008, Vancouver: Trafford Publishing. ISBN 978-1-4251-4132-5.

111. Herndon, J.M., *Uniqueness of Herndon's georeactor: Energy source and production mechanism for Earth's magnetic field.* https://arxiv.org/abs/0901.4509

112. Herndon, J.M., *Reevaporation of condensed matter during the formation of the solar system.* Proc. R. Soc. Lond, 1978. **A363**: p. 283-288.

113. Herndon, J.M., *Corruption of Science in America*, in *The Dot Connector*. 2011. http://www.nuclearplanet.com/corruption.pdf

114. Corredoira, M.L. and C.C. Perelman, eds. *Against the Tide: A Critical Review by Scientists of How Physics & Astronomy Get Done.* 2008, Universal Publishers: Boca Raton, Florida, USA. 265.

115. Herndon, J.M., *Some reflections on science and discovery.* Curr. Sci., 2015. **108**(11): p. 1967-1968.

116. Wegener, A.L., *Die Entstehung der Kontinente.* Geol. Rundschau, 1912. **3**: p. 276-292.

117. Vine, F.J. and D.H. Matthews, *Magnetic anomalies over oceanic ridges.* Nature, 1963. **199**: p. 947-949.

118. Hilgenberg, O.C., *Vom wachsenden Erdball.* 1933, Berlin: Giessmann and Bartsch. 56.

119. Carey, S.W., *The Expanding Earth.* 1976, Amsterdam: Elsevier. 488.

120. Beck, A.E., *Energy requirements in terrestrial expansion.* J. Geophys. Res., 1961. **66**: p. 1485-1490.

121. Cook, M.A. and A.J. Eardley, *Energy requirements in terrestrial expansion.* J. Geophys. Res., 1961. **66**: p. 3907-3912.

122. Herndon, J.M., *Solar System processes underlying planetary formation, geodynamics, and the georeactor.* Earth, Moon, and Planets, 2006. **99**(1): p. 53-99.

123. Herndon, J.M., *Energy for geodynamics: Mantle decompression thermal tsunami.* Curr. Sci., 2006. **90**(12): p. 1605-1606.

124. Herndon, J.M., *Indivisible Earth: Consequences of Earth's Early Formation as a Jupiter-Like Gas Giant*, L. Margulis, Editor. 2012, Thinker Media, Inc.

125. Frank, L.A., *Atmospheric holes and small comets.* Rev. Geophys., 1993. **31**(1): p. 1-28.

126. Frank, L.A., *The Big Splash*. 1990, New York: Birch Lane Press.

127. Gutzmer, J., et al., *Ancient sub-seafloor alteration of basaltic andesites of the Ongeluk Formation, South Africa: implications for the chemistry of Paleoproterozoic seawater.* Chemical Geology, 2003. **201**(1-2): p. 37-53.

128. Sparks, R., et al., *Dynamical constraints on kimberlite volcanism.* Journal of Volcanology and Geothermal Research, 2006. **155**(1-2): p. 18-48.

129. Herndon, J.M., *Origin of mountains and primary initiation of submarine canyons: the consequences of Earth's early formation as a Jupiter-like gas giant.* Curr. Sci., 2012. **102**(10): p. 1370-1372.

130. Dana, J.D., *On some results of the Earth's contraction from cooling including a discussion of the origin of mountains and the nature of the Earth's interior.* American Journal of Science, 1873. **3**(30): p. 423-443.

131. Le Conte, J., *On the structure and origin of mountains, with special reference to recent objections to the" contractional theory."*. American Journal of Science, 1878. **3**(92): p. 95-112.

132. Kossmat, F., *An English Translation of Palaögeographie-Geologische Geschichte Der Meere und Festländer by Franz Kossmat (1924)*. 2011: Edwin Mellen Press.

133. Ward, P.D., et al., *Measurements of the Cretaceous paleolatitude of Vancouver Island: consistent with the Baja-British Columbia hypothesis*. Science, 1997. **277**(5332): p. 1642-1645.

134. Herndon, J.M., *Potentially significant source of error in magnetic paleolatitude determinations*. Curr. Sci., 2011. **101**(3): p. 277-278.

135. Bijwaard, H. and W. Spakman, *Tomographic evidence for a narrow whole mantle plume below Iceland*. Earth Planet. Sci. Lett., 1999. **166**: p. 121-126.

136. Nataf, H.-C., *Seismic Imaging of Mantle Plumes*. Ann. Rev. Earth Planet. Sci., 2000. **28**: p. 391-417.

137. Hilton, D.R., et al., *Extreme He-3/He-4 ratios in northwest Iceland: constraining the common component in mantle plumes*. Earth Planet. Sci. Lett., 1999. **173**(1-2): p. 53-60.

138. Basu, A.R., et al., *High-^{3}He plume origin and temporal-spacial evolution of the Siberian flood basalts*. Sci., 1995. **269**: p. 882-825.

139. Basu, A.R., et al., *Early and late alkali igneous pulses and a high-^{3}He plume origin for the Deccan flood basalts*. Sci., 1993. **261**: p. 902-906.

140. Marty, B., et al., *He, Ar, Nd and Pb isotopes in volcanic rocks from Afar*. Geochem. J., 1993. **27**: p. 219-228.

141. Craig, H., et al., *Helium isotope ratios in Yellowstone and Lassen Park volcanic gases*. Geophysical Research Letters, 1978. **5**(11): p. 897-900.

142. Lowenstern, J.B., R.B. Smith, and D.P. Hill, *Monitoring super-volcanoes: geophysical and geochemical signals at Yellowstone and other large caldera systems.* Philosophical Transactions of the Royal Society A: Mathematical, Physical and Engineering Sciences, 2006. **364**(1845): p. 2055-2072.

143. Lowenstern, J.B. and S. Hurwitz, *Monitoring a supervolcano in repose: Heat and volatile flux at the Yellowstone Caldera.* Elements, 2008. **4**(1): p. 35-40.

144. Smith, R.B., et al., *Geodynamics of the Yellowstone hotspot and mantle plume: Seismic and GPS imaging, kinematics, and mantle flow.* Journal of Volcanology and Geothermal Research, 2009. **188**(1-3): p. 26-56.

145. Wotzlaw, J.-F., et al., *Linking rapid magma reservoir assembly and eruption trigger mechanisms at evolved Yellowstone-type supervolcanoes.* Geology, 2014. **42**(9): p. 807-810.

146. Herndon, J.M., *Impact of recent discoveries on petroleum and natural gas exploration: Emphasis on India.* Curr. Sci., 2010. **98**(6): p. 772-779.

147. Herndon, J.M., *New concept on the origin of petroleum and natural gas deposits.* J Petrol Explor Prod Technol 2017. **7**(2): p. 345-352.

148. Herndon, J.M., *Enhanced prognosis for abiotic natural gas and petroleum resources.* Curr. Sci., 2006. **91**(5): p. 596-598.

149. Blanchon, P. and J. Shaw, *Reef drowning during the last deglaciation: evidence for catastrophic sea-level rise and ice-sheet collapse.* Geology, 1995. **23**(1): p. 4-8.

150. Hagiwara, Y., *Geocatastrophe Mass Extinction and Geomagnetic Reversal.* Journal of Geography (Chigaku Zasshi), 1991. **100**(7): p. 1059-1076.

151. Kennett, J.P. and N. Watkins, *Geomagnetic polarity change, volcanic maxima and faunal extinction in the South Pacific.* Nature, 1970. **227**(5261): p. 930-934.

152. Irvine, T.N., *A global convection framework; concepts of symmetry, stratification, and system in the Earth's dynamic structure.* Economic Geology, 1989. **84**(8): p. 2059-2114.

153. Marzocchi, W. and F. Mulargia, *Feasibility of a synchronized correlation between Hawaiian hot spot volcanism and geomagnetiC polarity.* Geophysical Research Letters, 1990. **17**(8): p. 1113-1116.

154. Marzocchi, W., F. Mulargia, and P. Paruolo, *The correlation of geomagnetic reversals and mean sea level in the last 150 my.* Earth and planetary science letters, 1992. **111**(2-4): p. 383-393.

155. Hsu, K.J., *The great dying*. 1988: Ballantine Books.

156. Raup, D.M., *Magnetic reversals and mass extinctions.* Nature, 1985. **314**(6009): p. 341-343.

157. Raup, D.M. and J.J. Sepkoski, *Periodicity of extinctions in the geologic past.* Proceedings of the National Academy of Sciences, 1984. **81**(3): p. 801-805.

158. Hallam, A. and P. Wignall, *Mass extinctions and sea-level changes.* Earth-Science Reviews, 1999. **48**(4): p. 217-250.

159. Hallam, A., *Phanerozoic sea-level changes*. 1992: Columbia University Press.

160. Miall, A.D., *Exxon global cycle chart: An event for every occasion?* Geology, 1992. **20**(9): p. 787-790.

161. Miller, K.G., et al., *A 180-million-year record of sea level and ice volume variations from continental margin and deep-sea isotopic records.* Oceanography, 2011. **24**(2): p. 40-53.

162. Raup, D.M. and J.J. Sepkoski, *Mass extinctions in the marine fossil record.* Science, 1982. **215**(4539): p. 1501-1503.

163. Rohde, R.A. and R.A. Muller, *Cycles in fossil diversity.* Nature, 2005. **434**(7030): p. 208-210.

164. Chamberlin, T.C., *A group of hypotheses bearing on climatic changes.* The journal of geology, 1897. **5**(7): p. 653-683.

165. Moulton, F., *An attempt to test the nebular hypothesis by an appeal to the laws of dynamics.* The Astrophysical Journal, 1900. **11**: p. 103.

166. Chamberlin, T. and F. Moulton, *The development of the planetesimal hypothesis.* Science, 1909. **30**(775): p. 642-645.

167. Cameron, A.G.W., *Formation of the solar nebula.* Icarus, 1963. **1**: p. 339-342.

168. Goldrich, P. and W.R. Ward, *The formation of planetesimals.* Astrophys J., 1973. **183**(3): p. 1051-1061.

169. Chambers, J. and G. Wetherill, *Making the terrestrial planets: N-body integrations of planetary embryos in three dimensions.* Icarus, 1998. **136**(2): p. 304-327.

170. Larimer, J.W., *Chemistry of the solar nebula.* Space Sci. Rev., 1973. **15**(1): p. 103-119.

171. Mehta, A.V., *The role of vortices in the formation of the solar system.* 1998, Massachusetts Institute of Technology.

172. Bell, J.F., *Water on planets.* Proceedings of the International Astronomical Union, 2009. **5**(H15): p. 29-44.

173. Leigh, C., *The detection and characterisation of extrasolar planets.* 2004, University of St Andrews.

174. Grossman, L., *Condensation in the primitive solar nebula.* Geochim. Cosmochim. Acta, 1972. **36**: p. 597-619.

175. Elkins-Tanton, L.T., *Magma oceans in the inner solar system.* Annual Review of Earth and Planetary Sciences, 2012. **40**: p. 113-139.

176. Siebert, J., et al., *Metal–silicate partitioning of Ni and Co in a deep magma ocean.* Earth and Planetary Science Letters, 2012. **321**: p. 189-197.

177. Murray, N., et al., *Migrating planets.* Science, 1998. **279**(5347): p. 69-72.

178. Herndon, J.M., *Evidence contrary to the existing exo-planet migration concept.* https://arxiv.org/abs/astro-ph/0612726

179. Toci, C., et al., *Planet migration, resonant locking, and accretion streams in PDS 70: comparing models and data.* Monthly Notices of the Royal Astronomical Society, 2020. **499**(2): p. 2015-2027.

180. Mojzsis, S.J., et al., *Onset of giant planet migration before 4480 million years ago.* The Astrophysical Journal, 2019. **881**(1): p. 44.

181. Mason, B., *The classification of chondritic meteorites.* Amer. Museum Novitates, 1962. **2085**: p. 1-20.

182. Suess, H.E. and H.C. Urey, *Abundances of the elements.* Rev. Mod. Phys., 1956. **28**: p. 53-74.

183. Anders, E. and N. Grevesse, *Abundances of the elements: Meteoritic and solar.* Geochimica et Cosmochimica acta, 1989. **53**(1): p. 197-214.

184. Herndon, J.M. and H.E. Suess, *Can the ordinary chondrites have condensed from a gas phase?* Geochim. Cosmochim. Acta, 1977. **41**: p. 233-236.

185. Herndon, J.M., *Discovery of fundamental mass ratio relationships of whole-rock chondritic major elements: Implications on ordinary chondrite formation and on planet Mercury's composition.* Curr. Sci., 2007. **93**(3): p. 394-398.

186. Larson, E., et al., *Thermomagnetic analysis of meteorites, 1. C1 chondrites.* Earth and Planetary Science Letters, 1974. **21**(4): p. 345-350.

187. Hyman, M., R. MW, and H. JM, *Magnetite heterogeneity among Cl chondrites.* Geochemical Journal, 1979. **13**(1): p. 37-39.

188. Kant, I., *Allgemeine Naturgeschichte und Theorie des Himmels (Universal natural history and theory of the heavens).* Trans. by Ian Johnston. Arlington, VA: Richer Resources, 1755.

189. Laplace, P.S.d. *Pierre Simon de Laplace.* in *Exposition du système du monde.* 1796.

190. Eucken, A., *Physikalisch-chemische Betrachtungen ueber die frueheste Entwicklungsgeschichte der Erde.* Nachr. Akad. Wiss. Goettingen, Math.-Kl., 1944: p. 1-25.

191. Kuiper, G.P., *On the evolution of the protoplanets.* Proc. Nat. Acad. Sci. USA, 1951. **37**: p. 383-393.

192. Urey, H.C., *On the Dissipation of Gas and Volatilized Elements from Protoplanets.* The Astrophysical Journal Supplement Series, 1954. **1**: p. 147.

193. Herndon, J.M., *Composition of the deep interior of the earth: divergent geophysical development with fundamentally different geophysical implications.* Phys. Earth Plan. Inter, 1998. **105**: p. 1-4.

194. Blewett, D.T., et al., *Hollows on Mercury: MESSENGER Evidence for Geologically Recent Volatile-Related Activity.* Science, 2011. **333**: p. 1859-1859.

195. Foster, R.G. and T. Roenneberg, *Human responses to the geophysical daily, annual and lunar cycles.* Current biology, 2008. **18**(17): p. R784-R794.

196. Andreatta, G. and K. Tessmar-Raible, *The still dark side of the moon: molecular mechanisms of lunar-controlled rhythms and clocks.* Journal of molecular biology, 2020. **432**(12): p. 3525-3546.

197. Aeberhard, A. and S. Rist, *Transdisciplinary co-production of knowledge in the development of organic agriculture in Switzerland.* Ecological Economics, 2009. **68**(4): p. 1171-1181.

198. Irangani, M. and Y. Shiratake, *Indigenous techniques used in rice cultivation in Sri Lanka: An analysis from an agricultural history perspective.* 2013.

199. Varisco, D.M., *The agricultural marker stars in Yemeni folklore.* Asian Folklore Studies, 1993: p. 119-142.

200. Fox, H.M., *Lunar periodicity in reproduction.* Proceedings of the Royal Society of London. Series B, Containing Papers of a Biological Character, 1924. **95**(671): p. 523-550.

201. Morgan, E. and G. Harris. *The role of tidal activity rhythms in the migrations of an estuarine amphipod.* in *Behavioural Rhythms, Readings from the 19th International Ethological Conference, IEC Universite'Paul Sabatier/Toulouse.* 1986.

202. Telfer, T.C., et al., *Attraction of Hawaiian seabirds to lights: conservation efforts and effects of moon phase.* Wildlife Society Bulletin (1973-2006), 1987. **15**(3): p. 406-413.

203. Pérez-Granados, C., K.-L. Schuchmann, and M.I. Marques, *Addicted to the moon: vocal output and diel pattern of vocal activity in two Neotropical nightjars is related to moon phase.* Ethology Ecology & Evolution, 2022. **34**(1): p. 66-81.

204. Pop, V. *Lunar exploration and the social dimension.* in *Earth-like Planets and Moons.* 2002.

205. Huxley, A., *Meditation on the Moon.* Music at Night, 1950.

206. Wilson, E.W., *The moon and the American Indian.* Western Folklore, 1965. **24**(2): p. 87-100.

207. Markey, A., *Selene: Lady Mount Cashell's Lunar Utopia.* Women's Writing, 2014. **21**(4): p. 559-574.

208. Greenleaf, C., *Moon Spell Magic: Invocations, Incantations & Lunar Lore for a Happy Life.* 2017: Mango Media Inc.

209. Waltz, S.C. *In Defense of Moonlight.* in *Beethoven Forum.* 2007. University of Illinois Press Champaign.

210. Zheng, Y., et al., *China's lunar exploration program: present and future.* Planetary and Space Science, 2008. **56**(7): p. 881-886.

211. Li, H. *On the Image of Full Moon in This Lunar Beauty by WH Auden.* in *2015 International Conference on Social Science, Education Management and Sports Education.* 2015. Atlantis Press.

212. *https://en.wikipedia.org/wiki/Luna_3*

213. Crawford, I.A., *Lunar resources: A review.* Progress in Physical Geography, 2015. **39**(2): p. 137-167.

214. Jaumann, R., et al., *Geology, geochemistry, and geophysics of the Moon: Status of current understanding.* Planetary and Space Science, 2012. **74**(1): p. 15-41.

215. Herschel, W., *III. On the nature and construction of the sun and fixed stars.* Philosophical Transactions of the Royal Society of London, 1795(85): p. 46-72.

216. Anderson, W., *The Philosophy of Ancient Greece Investigated: In Its Origin and Progress.* 1791: Smellie.

217. Urey, H.C., *The Planets.* 1952, New Haven: Yale University Press.

218. *https://en.wikipedia.org/wiki/Dorion_Sagan*

219. *https://en.wikipedia.org/wiki/Carl_Sagan*

220. *https://en.wikipedia.org/wiki/Lynn_Margulis*

221. Zhu, M.H., et al., *Are the Moon's nearside-farside asymmetries the result of a giant impact?* Journal of Geophysical Research: Planets, 2019. **124**(8): p. 2117-2140.

222. Head, J.W. and A. Gifford, *Lunar mare domes: Classification and modes of origin.* The moon and the planets, 1980. **22**(2): p. 235-258.

223. Schultz, P. and P. Spudis, *Beginning and end of lunar mare volcanism.* Nature, 1983. **302**(5905): p. 233-236.

224. Herndon, J.M., *New explanation for the near-side/far-side lunar maria disparity.* Journal of Geography, Environment and Earth Science International, 2022. **26**(1): p. 1-4.

225. Herndon, J.M., *Making sense of chondritic meteorites.* Advances in Social Sciences Research Journal, 2022. **9**(2): p. 82-102.

226. Herndon, J.M., *Energy for geodynamics: Mantle decompression thermal tsunami.* Curr. Sci., 2006. **90**(12): p. 1605-1606.

227. Yoder, J., H. S., *The great basalt 'floods'.* S. Afr. Tydskr. Geol., 1988. **91**(2): p. 139-156.

228. Mitchell, C. and M. Widdowson, *A geological map of the southern Deccan Traps, India and its structural implications.* Journal of the Geological Society, 1991. **148**(3): p. 495-505.

229. Basu, A.R., A. Saha-Yannopoulos, and P. Chakrabarty, *A precise geochemical volcano-stratigraphy of the Deccan traps.* Lithos, 2020. **376**: p. 105754.

230. Renne, P.R. and A.R. Basu, *Rapid eruption of the Siberian Traps flood basalts at the Permo-Triassic boundary.* Science, 1991. **253**(5016): p. 176-179.

231. Bagdasaryan, T.E., et al., *Thermal history of the Siberian Traps Large Igneous Province revealed by new thermochronology data from intrusions.* Tectonophysics, 2022. **836**: p. 229385.

232. Hilton, D.R. and D. Porcelli, *Noble gases as mantle tracers.*, in *The Mantle and Core*, R.W. Carlson, Editor. 2003, Elsevier-Pergamon: Oxford. p. 277-318.

233. Runcorn, S., *An ancient lunar magnetic dipole field.* Nature, 1975. **253**(5494): p. 701-703.

234. Garrick-Bethell, I., et al., *Further evidence for early lunar magnetism from troctolite 76535.* Journal of Geophysical Research: Planets, 2017. **122**(1): p. 76-93.

235. Herndon, J.M., *Paradigm Shifts: A Primer for Students, Teachers, Scientists and the Curious.* 2021: Amazon.com.

236. Herndon, J.M., *What's wrong with this picture?* Advances in Social Sciences Research Journal, 2022. **9**(3): p. 64-69.

237. Vermillion, R.E., *On the center-of-mass offset of the moon.* American Journal of Physics, 1976. **44**(10): p. 1014.

238. Keeton, C., *Tidal Forces,* in *Principles of Astrophysics.* 2014, Springer. p. 79-88.

239. Yanchukovsky, V., *Solar activity and Earth seismicity.* Solar-Terrestrial Physics, 2021. **7**(1): p. 67-77.

240. Semeida, M., et al., *Examination of the relationship between solar activity and earth seismicity during the weak solar cycle 23.* Bulgarian Academy of Sciences ISSN 1313–0927: p. 5.

241. Ulukavak, M. and S. Inyurt, *Seismo-ionospheric precursors of strong sequential earthquakes in Nepal region.* Acta Astronautica, 2020. **166**: p. 123-130.

242. Khegai, V., et al., *Solar Activity, Galactic Cosmic Ray Variations, and the Global Seismicity of the Earth.* Geomagnetism and Aeronomy, 2021. **61**(1): p. S36-S47.

243. Novikov, V., et al., *Space weather and earthquakes: possible triggering of seismic activity by strong solar flares.* Annals of Geophysics, 2020. **63**(5): p. PA554-PA554.

244. Gonzalez-Esparza, J., et al., *Space weather events, hurricanes, and earthquakes in Mexico in September 2017.* Space Weather, 2018. **16**(12): p. 2038-2051.

245. Nurtaev, B., *General Relativity Theory and Earthquakes.* Journal of the Georgian Geophysical Society, 2020. **23**(1).

246. Anagnostopoulos, G., et al., *The sun as a significant agent provoking earthquakes.* The European Physical Journal Special Topics, 2021. **230**(1): p. 287-333.

247. Vasilieva, I. and V.V. Zharkova, *Terrestrial volcanic eruptions and their link with solar activity.* https://solargsm.com

248. Ma, L., Z. Yin, and Y. Han, *Possible Influence of Solar Activity on Global Volcanicity.* Earth Science Research, 2018. **7**(110): p. 10.5539.

249. Weidenschilling, S.J., *Formation of planetesimals and accretion of the terrestrial planets.* Space Science Reviews, 2000. **92**(1): p. 295-310.

250. Glatzmaier, G.A. and P. Olson, *Probing the geodynamo.* Scientific American, 2005. **292**(4): p. 50-57.

251. Chandrasekhar, S., *Thermal Convection.* Proc. Amer. Acad. Arts Sci., 1957. **86**(4): p. 323-339.

252. Dziewonski, A.M. and D.A. Anderson, *Preliminary reference Earth model.* Phys. Earth Planet. Inter., 1981. **25**: p. 297-356.

253. Jacobs, J., *The cause of superchrons.* Astronomy & Geophysics, 2001. **42**(6): p. 6.30-6.31.

254. Driscoll, P.E. and D.A. Evans, *Frequency of Proterozoic geomagnetic superchrons.* Earth and Planetary Science Letters, 2016. **437**: p. 9-14.

255. Herndon, J.M., *NASA: Politics Above Science.* Amazon Kindle Direct Publishing, 2018.

256. Connerney, J., et al., *The global magnetic field of Mars and implications for crustal evolution.* Geophysical Research Letters, 2001. **28**(21): p. 4015-4018.

257. Lillis, R.J., et al., *An improved crustal magnetic field map of Mars from electron reflectometry: Highland volcano magmatic history*

and the end of the Martian dynamo. Icarus, 2008. **194**(2): p. 575-596.

258. Lillis, R.J., et al., *Time history of the Martian dynamo from crater magnetic field analysis.* Journal of Geophysical Research: Planets, 2013. **118**(7): p. 1488-1511.

259. Schultz, R.A., *Structural development of coprates chasma and western ophir planum, valles marineris rift, mars.* Journal of Geophysical Research: Planets, 1991. **96**(E5): p. 22777-22792.

260. Andrews-Hanna, J.C., *The formation of Valles Marineris: 2. Stress focusing along the buried dichotomy boundary.* Journal of Geophysical Research: Planets, 2012. **117**(E4).

261. Davis, J.M., et al., *A record of syn-tectonic sedimentation revealed by perched alluvial fan deposits in Valles Marineris, Mars.* Geology, 2021. **49**(10): p. 1250-1254.

262. Orosei, R., et al., *The global search for liquid water on Mars from orbit: Current and future perspectives.* Life, 2020. **10**(8): p. 120.

263. Oehler, D.Z. and G. Etiope, *Methane seepage on Mars: where to look and why.* Astrobiology, 2017. **17**(12): p. 1233-1264.

264. Herndon, J.M., *Herndon's Earth and the Dark Side of Science.* Amazon Kindle Direct Publishing, 2014.

265. Masaitis, V.L., *Permian and Triassic volcanism of Siberia.* Zapiski VMO, 1983. **part CXII**(4): p. 412-425.

266. Horn, M.K., *Giant Fields 1869-2003*, in *Giant Oil and Gas Fields of the Decade, 1990-1999*, M.K. Halbouty, Editor. 2003, AAPG: Houston.

267. Benfield, A.F., *Terrestrial heat flow in Great Britain.* Proc. R. Soc. Lond, 1939. **Ser A 173**: p. 428-450.

268. Bullard, E.C., *Heat flow in South Africa.* Proc. R. Soc. Lond, 1939. **Ser. A 173**: p. 474-502.

269. Revelle, R. and A.E. Maxwell, *Heat flow through the floor of the eastern North Pacific Ocean.* Nature, 1952. **170**: p. 199-200.

270. Di Achille, G. and B.M. Hynek, *Ancient ocean on Mars supported by global distribution of deltas and valleys.* Nature Geoscience, 2010. **3**(7): p. 459-463.

271. Nazari-Sharabian, M., et al., *Water on Mars—A Literature Review.* Galaxies, 2020. **8**(2): p. 40.

272. Kass, D. and Y.L. Yung, *Loss of atmosphere from Mars due to solar wind-induced sputtering.* Science, 1995. **268**(5211): p. 697-699.

273. Von Hagen, V.W., *The ancient sun kingdoms of the Americas.* Vol. 1. 2017: Pickle Partners Publishing.

274. Benson, E.P. and A.G. Cook, *Ritual sacrifice in ancient Peru.* 2001: University of Texas Press.

275. Breasted, J.H., *Development of religion and thought in ancient Egypt.* Vol. 45. 1972: University of Pennsylvania Press.

276. Herndon, J.M., *New concept for internal heat production in hot Jupiter exo-planets, thermonuclear ignition of dark galaxies, and the basis for galactic luminous star distributions.* Curr. Sci., 2009. **96**: p. 1453-1456.

277. Bacquerel, H., *Sur les radiations emises par phosphorescence.* Comptes Rendus, 1896. **122**: p. 501-503.

278. Malley, M.C., *Radioactivity: a history of a mysterious science.* 2011: Oxford University Press.

279. Oliphant, M.L., P. Harteck, and E. Rutherford, *Transmutation effects observed with heavy hydrogen.* Nature, 1934. **133**: p. 413.

280. Gamow, G. and E. Teller, *The rate of selective thermonuclear reactions.* Phys. Rev., 1938. **53**: p. 608-609.

281. Bethe, H.A., *Energy production in stars.* Phys. Rev., 1939. **55**(5): p. 434-456.

282. Hayashi, C. and T. Nakano, *Thermal and dynamic properties of a protostar and its contraction to the stage of quasi-static equilibrium.* Prog. theor. Physics, 1965. **35**: p. 754-775.

283. Larson, R.B., *Gravitational torques and star formation.* Mon. Not. R. astr. Soc., 1984. **206**: p. 197-207.

284. Stahler, S.W., et al., *The early evolution of protostellar disks.* Astrophys. J., 1994. **431**: p. 341-358.

285. Hahn, O. and F. Strassmann, *Uber den Nachweis und das Verhalten der bei der Bestrahlung des Urans mittels Neutronen entstehenden Erdalkalimetalle.* Die Naturwissenschaften, 1939. **27**: p. 11-15.

286. Serber, R., *The Los Alamos primer: The first lectures on how to build an atomic bomb.* 1992: Univ of California Press.

287. Rhodes, R., *The Making of the Atomic Bomb.* 1986, New York: Simon & Schuster.

288. Sekimoto, H., *Nuclear reactor theory.* 2008: COE-INES, Tokyo Institute of Technology.

289. Rhodes, R., *Dark Sun: The Making of the Hydrogen Bomb.* Vol. 2. 1996: Simon and Schuster.

290. Kuroda, P.K., *On the nuclear physical stability of the uranium minerals.* J. Chem. Phys., 1956. **25**(4): p. 781-782.

291. Kuroda, P.K., *On the infinite multiplication constant and the age of uranium minerals.* J. Chem. Phys., 1956. **25**(6): p. 1295-1296.

292. Frejacques, C., et al., in *The Oklo Phenomenon.* 1975, I.A.E.A.: Vienna. p. 509.

293. Hagemann, R., et al., in *The Oklo Phenomenon.* 1975, I.A.E.A.: Vienna. p. 415.

294. Conrath, B.J., et al., in *Uranus*, J.T. Bergstralh, E.D. Miner, and M.S. Mathews, Editors. 1991, University of Arizona Press: Tucson.

295. Hubbard, W.B., *Interiors of the giant planets*, in *The New Solar System*, A. Chaikin, Editor. 1990, Sky Publishing Corp.: Cambridge, MA. p. 134-135.

296. Stevenson, J.D., *The outer planets and their satellites*, in *The Origin of the Solar System*, S.F. Dermott, Editor. 1978, Wiley: New York. p. 395-431.

297. Herndon, J.M., *NASA: Politics Above Science*. Amazon Kindle Direct Publishing, 2018.

298. Rubin, V.C., *The rotation of spiral galaxies.* Science, 1983. **220**: p. 1339-1344.

299. Arun, K., S. Gudennavar, and C. Sivaram, *Dark matter, dark energy, and alternate models: A review.* Advances in Space Research, 2017. **60**(1): p. 166-186.

300. Burbidge, E.M., et al., *Synthesis of the elements in stars.* Rev. Mod. Phys., 1957. **29**(4): p. 547-650.

301. Hubble, E., *A relation between distance and radial velocity among extra-galactic nebulae.* Proc. Nat. Acad. Sci. USA, 1929. **15**: p. 168-173.

302. Slipher, V.M., *The radial velocity of the Andromeda Nebula.* Lowell Observatory Bulletin, 1913. **2**: p. 56-57.

303. Penzias, A.A. and R.W. Wilson, *A measurement of excess antenna temperature at 4080 Mc/s.* Astrophys. J., 1965. **142**: p. 419-421.

304. Ceballos, G., P.R. Ehrlich, and R. Dirzo, *Biological annihilation via the ongoing sixth mass extinction signaled by vertebrate population losses and declines.* Proceedings of the National Academy of Sciences, 2017. **114**(30): p. E6089-E6096.

305. Blanchard, J., *Living Planet Report 2020: Bending the Curve of Biodiversity Loss.* 2020.

306. Dirzo, R., et al., *Defaunation in the Anthropocene.* Science, 2014. **345**(6195): p. 401-406.

307. Bradshaw, C.J., et al., *Underestimating the challenges of avoiding a ghastly future.* Frontiers in Conservation Science, 2021. **1**: p. 9.

308. Carson, R.L., *Silent Spring.* 1962, Boston, MA: Houghton Mifflin.

309. Seneff, S., *Toxic Legacy: How the Weedkiller Glyphosate Is Destroying Our Health and the Environment.* 2021: Chelsea Green Publishing.

310. Shearer, C., et al., *Quantifying expert consensus against the existence of a secret large-scale atmospheric spraying program.* Environ. Res. Lett., 2016. **11**(8): p. p. 084011.

311. Tingley, D. and G. Wagner, *Solar geoengineering and the chemtrails conspiracy on social media.* Palgrave Communications, 2017. **3**(1): p. 12.

312. Lovelock, J. and L. Margulis, *The Gaia Hypothesis.* 2007, New York.

313. Lovelock, J.E. and L. Margulis, *Atmospheric homeostasis by and for the biosphere: the Gaia hypothesis.* Tellus, 1974. **26**(1-2): p. 2-10.

314. Margulis, L. and J.E. Lovelock, *Biological modulation of the Earth's atmosphere.* Icarus, 1974. **21**(4): p. 471-489.

315. Harrop, S.R., *'Living in harmony with nature'? Outcomes of the 2010 Nagoya Conference of the Convention on Biological Diversity.* Journal of Environmental Law, 2011. **23**(1): p. 117-128.

316. Sólon, P., *The rights of mother earth.* The climate crisis. South African and global democratic eco-socialist alternatives, 2018: p. 107-130.

317. McGregor, D., *Mother Earth.* An Insider's Guide to a Rapidly Changing Planet, 2020: p. 133.

318. Herndon, J.M. and M. Whiteside, *Geophysical consequences of tropospheric particulate heating: Further evidence that anthropogenic global warming is principally caused by particulate pollution.* Journal of Geography, Environment and Earth Science International, 2019. **22**(4): p. 1-23.

319. http://www.nuclearplanet.com/usaf1.pdf

320. Herndon, J.M. and M. Whiteside, *California wildfires: Role of undisclosed atmospheric manipulation and geoengineering.* J. Geog. Environ. Earth Sci. Intn., 2018. **17**(3): p. 1-18.

321. Herndon, J.M., R.D. Hoisington, and M. Whiteside, *Chemtrails are not contrails: Radiometric evidence.* J. Geog. Environ. Earth Sci. Intn., 2020. **24**(2): p. 22-29.

322. Hagen, M., et al., *Geoengineering disinformation: Two opposing testimonies and the stakes for humanity.* Advances in Social Sciences Research Journal, 2024. **11**(5): p. 254-266.

323. Borm, P.J.A., *Toxicity and occupational health hazards of coal fly ash (cfa). A review of data and comparison to coal mine dust.* Ann. occup. Hyg., 1997. **41**(6): p. 659-676.

324. Roy, W.R., R. Thiery, and J.J. Suloway, *Coal fly ash: a review of the literature and proposed classification system with emphasis on environmental impacts.* Environ. Geology Notes #96, 1981.

325. Walls, S.J., et al., *Ecological risk assessment for residual coal fly ash at Watts Bar Reservoir, Tennessee: Site setting and problem formulation.* Integrated environmental assessment and management, 2015. **11**(1): p. 32-42.

326. Dwivedi, A. and M.K. Jain, *Fly ash–waste management and overview: A Review.* Recent Research in Science and Technology, 2014. **6**(1).

327. Moreno, N., et al., *Physico-chemical characteristics of European pulverized coal combustion fly ashes.* Fuel, 2005. **84**: p. 1351-1363.

328. Jigyasu, D.K., et al., *High mobility of aluminum in Gomati River Basin: implications to human health.* Curr. Sci., 2015. **108**(3): p. 434-438.

329. Herndon, J.M., *Aluminum poisoning of humanity and Earth's biota by clandestine geoengineering activity: implications for India.* Curr. Sci., 2015. **108**(12): p. 2173-2177.

330. *http://www.nuclearplanet.com/Retraction_Deception.html*

331. Herndon, J.M. and M. Whiteside, *Nature as a Weapon of Global War: The Deliberate Destruction of Life on Earth.* Worldwide: Amazon Kindle Direct Publishing, 2021.

332. Herndon, J.M. and M. Whiteside, *Chemtrails are not Contrails: The Face of Evil.* Amazon Kindle Direct Publishing, 2022.

333. Herndon, J.M., D.D. Williams, and M. Whiteside, *Previously unrecognized primary factors in the demise of endangered torrey pines: A microcosm of global forest die-offs.* J. Geog. Environ. Earth Sci. Intn., 2018. **16**(4): p. 1-14.

334. Suloway, J.J., et al., *Chemical and toxicological properties of coal fly ash*, in *Environmental Geology Notes 105.* 1983, Illinois Department of Energy and Natural Resources: Illinois.

335. Whiteside, M. and J.M. Herndon, *Aerosolized coal fly ash: A previously unrecognized primary factor in the catastrophic global demise of bird populations and species.* Asian J. Biol., 2018. **6**(4): p. 1-13.

336. Herndon, J.M. and M. Whiteside, *Contamination of the biosphere with mercury: Another potential consequence of on-going climate manipulation using aerosolized coal fly ash* J. Geog. Environ. Earth Sci. Intn., 2017. **13**(1): p. 1-11.

337. Herndon, J.M. and M. Whiteside, *Aerosolized coal fly ash particles, the main cause of stratospheric ozone depletion, not chlorofluorocarbon gases.* European Journal of Applied Sciences, 2022. **10**(3): p. 586-603.

338. Whiteside, M. and J.M. Herndon, *Destruction of stratospheric ozone: Role of aerosolized coal fly ash iron.* European Journal of Applied Sciences, 2022. **10**(4): p. 143-153.

339. Whiteside, M. and J.M. Herndon, *New paradigm: Coal fly ash as the main cause of stratospheric ozone depletion.* European Journal of Applied Sciences, 2022. **10**(5): p. 207-221.

340. Herndon, J.M., R.D. Hoisington, and M. Whiteside, *Deadly ultraviolet UV-C and UV-B penetration to Earth's surface: Human and environmental health implications.* J. Geog. Environ. Earth Sci. Intn., 2018. **14**(2): p. 1-11.

341. Whiteside, M. and J.M. Herndon, *Role of aerosolized coal fly ash in the global plankton imbalance: Case of Florida's toxic algae crisi.* Asian Journal of Biology, 2019. **8**(2): p. 1-24.

342. Whiteside, M. and J.M. Herndon, *Geoengineering, coal fly ash and the new heart-Iron connection: Universal exposure to iron oxide nanoparticulates.* Journal of Advances in Medicine and Medical Research, 2019. **31**(1): p. 1-20.

343. Whiteside, M. and J.M. Herndon, *Aerosolized coal fly ash: Risk factor for neurodegenerative disease.* Journal of Advances in Medicine and Medical Research, 2018. **25**(10): p. 1-11.

344. Maher, B., et al., *Iron-rich air pollution nanoparticles: An unrecognised environmental risk factor for myocardial mitochondrial dysfunction and cardiac oxidative stress.* Environmental research, 2020. **188**: p. 109816.

345. Whiteside, M. and J.M. Herndon, *Aerosolized coal fly ash: Risk factor for COPD and respiratory disease.* Journal of Advances in Medicine and Medical Research, 2018. **26**(7): p. 1-13.

346. Whiteside, M. and J.M. Herndon, *Coal fly ash aerosol: Risk factor for lung cancer.* Journal of Advances in Medicine and Medical Research, 2018. **25**(4): p. 1-10.

347. Herndon, J.M. and M. Whiteside, *Aerosol particulates, SARS-CoV-2, and the broader potential for global devastation.* Open Access Journal of Internal Medicine, 2020. **3**(1): p. 14-21.

348. Whiteside, M. and J.M. Herndon, *COVID-19, immunopathology, particulate pollution, and iron balance.* Journal of Advances in Medicine and Medical Research, 2020. **32**(18): p. 43-60.

349. Herndon, J.M., D.D. Williams, and M. Whiteside, *Previously unrecognized primary factors in the demise of endangered torrey pines: A microcosm of global forest die-offs.* J. Geog. Environ. Earth Sci. Intn. , 2018. **16**(4): p. 1-14.

350. Herndon, J.M., D.D. Williams, and M.W. Whiteside, *Ancient Giant Sequoias are dying: Scientists refuse to acknowledge the cause.* Advances in Social Sciences Research Journal, 2021. **8**(9): p. 57-70.

351. Herndon, J.M., *Adverse agricultural consequences of weather modification.* AGRIVITA Journal of agricultural science, 2016. **38**(3): p. 213-221.

352. Whiteside, M. and J.M. Herndon, *Previously unacknowledged potential factors in catastrophic bee and insect die-off arising from coal fly ash geoengineering* Asian J. Biol., 2018. **6**(4): p. 1-13.

353. Herndon, J.M. and M. Whiteside, *Unacknowledged potential factors in catastrophic bat die-off arising from coal fly ash geoengineering.* Asian Journal of Biology, 2019. **8**(4): p. 1-13.

354. Sharifan, H., *Alarming the impacts of the organic and inorganic UV blockers on endangered coral's species in the Persian Gulf: A scientific concern for coral protection.* Sustainable Futures, 2020. **2**: p. 100017.

355. MacDonald, G.J., *How to wreck the environment.* Unless Peace Comes: A Scientific Forecast of New Weapons, 1968: p. 181-205.

356. Staff, R. *Ahmadinejad says enemies destroy Iran's rain clouds - reports.* Commodity News, 2011.

357. http://cyprus-mail.com/2016/02/17/minister-pledges-probe-into-chemtrails/

358. Herndon, J.M., M. Whiteside, and I. Baldwin, *Fifty Years after "How to Wreck the Environment": Anthropogenic Extinction of Life on Earth.* J. Geog. Environ. Earth Sci. Intn., 2018. **16**(3): p. 1-15.

359. Herndon, J.M. and M. Whiteside, *Technology Bill of Rights needed to protect human and environmental health and the U. S. Constitutional Republic* Advances in Social Sciences Research Journal, 2020. **7**(6).

360. MacDonald, G.J., *How to wreck the environment*, in *Unless Peace Comes: A Scientific Forecast of New Weapons.* 1968, The Viking Press: New York. p. 181-205.

361. Herndon, J.M. and M. Whiteside, *Global Environmental Warfare.* Advances in Social Sciences Research Journal, 2020. **7**(4): p. 411-422.

362. Herndon, J.M., M. Whiteside, and I. Baldwin, *The ENMOD treaty and the sanctioned assault on agriculture and human and environmental health.* Agrotechnology, 2020. **9**(191): p. 1-9.

363. Herndon, J.M. and M. Whiteside, *Environmental warfare against American citizens: An open letter to the Joint Chiefs of Staff.* Advances in Social Sciences Research Journal, 2020. **7**(8): p. 382-397.

364. Herndon, J.M., M. Whiteside, and I. Baldwin, *Open letter to the International Criminal Court alleging United Nations complicity in planetary treason.* Advances in Social Sciences Research Journal, 2022. **9**(10): p. 243-258.

365. Herndon, J.M., *An indication of intentional efforts to cause global warming and glacier melting.* J. Geography Environ. Earth Sci. Int., 2017. **9**(1): p. 1-11.

366. Herndon, J.M., *Evidence of variable Earth-heat production, global non-anthropogenic climate change, and geoengineered global*

warming and polar melting. *J. Geog. Environ. Earth Sci. Intn.*, 2017. **10**(1): p. 16.

367. Herndon, J.M. and M. Whiteside, *Further evidence that particulate pollution is the principal cause of global warming: Humanitarian considerations.* Journal of Geography, Environment and Earth Science International, 2019. **21**(1): p. 1-11.

368. Herndon, J.M., *Scientific misrepresentation and the climate-science cartel.* J. Geog. Environ. Earth Sci. Intn., 2018. **18**(2): p. 1-13.

369. Herndon, J.M., *Fundamental climate science error: Concomitant harm to humanity and the environment* J. Geog. Environ. Earth Sci. Intn., 2018. **18**(3): p. 1-12.

370. Herndon, J.M., *Role of atmospheric convection in global warming.* J. Geog. Environ. Earth Sci. Intn., 2019. **19**(4): p. 1-8.

371. Herndon, J.M., *World War II holds the key to understanding global warming and the challenge facing science and society.* J. Geog. Environ. Earth Sci. Intn., 2019. **23**(4): p. 1-13.

372. Abdussamatov, H.I., *The Sun defines the climate.* Russian journal "Nauka i Zhizn" ("Science and Life"), 2008. **1**: p. 34-42.

373. Abdussamatov, H.I., *Grand minimum of the total solar irradiance leads to the little ice age.* Geol. Geosci., 2013. **2**(2): p. 1-10.

374.   https://www.ipcc.ch/site/assets/uploads/2018/02/WG1AR5_all_final.pdf

375. Costella, J., ed. *The Climategate Emails.* 2010, The Lavoisier Group: Australia.

376. Phalgune, A., et al. *Garbage in, garbage out? An empirical look at oracle mistakes by end-user programmers.* in *Visual Languages and Human-Centric Computing, 2005 IEEE Symposium on.* 2005. IEEE.

377. Curry, J.A. and P.J. Webster, *Climate science and the uncertainty monster.* Bulletin of the American Meteorological Society, 2011. **92**(12): p. 1667-1682.

378. Lovelock, J., *The Vanishing Face of Gaia: A Final Warning* 2009, London: Allen Lane/Penguine.

379. https://www.climate.gov/maps-data/primer/climate-forcing

380. Andreae, M.O., C.D. Jones, and P.M. Cox, *Strong present-day aerosol cooling implies a hot future.* Nature, 2005. **435**(7046): p. 1187.

381. Myhre, G., et al., *Anthropogenic and natural radiative forcing.* Climate Change, 2013. **423**: p. 658-740.

382. Bond, T.C. and H. Sun, *Can reducing black carbon emissions counteract global warming?* Environ. Sci. Technol., 2005. **39**: p. 5921-5926.

383. Letcher, T.M., *Why do we have global warming?*, in *Managing Global Warming.* 2019, Elsevier. p. 3-15.

384. Summerhayes, C.P. and J. Zalasiewicz, *Global warming and the Anthropocene.* Geology Today, 2018. **34**(5): p. 194-200.

385. Ångström, A., *On the atmospheric transmission of sun radiation and on dust in the air.* Geografiska Annaler, 1929. **11**(2): p. 156-166.

386. Robock, A., *Enhancement of surface cooling due to forest fire smoke.* Science, 1988: p. 911-913.

387. Robock, A., *Surface cooling due to forest fire smoke.* Journal of Geophysical Research: Atmospheres, 1991. **96**(D11): p. 20869-20878.

388. McCormick, R.A. and J.H. Ludwig, *Climate modification by atmospheric aerosols.* Science, 1967. **156**(3780): p. 1358-1359.

389. Andreae, M.O. and A. Gelencsér, *Black carbon or brown carbon? The nature of light-absorbing carbonaceous aerosols.* Atmospheric Chemistry and Physics, 2006. **6**(10): p. 3131-3148.

390. Wang, C., G.-R. Jeong, and N. Mahowald, *Particulate absorption of solar radiation: anthropogenic aerosols vs. dust.* Atmospheric Chemistry and Physics, 2009. **9**(12): p. 3935-3945.

391. Ramanathan, V. and G. Carmichael, *Global and regional climate changes due to black carbon.* Nature geoscience, 2008. **1**(4): p. 221.

392. Fan, J., et al., *Potential aerosol indirect effects on atmospheric circulation and radiative forcing through deep convection.* Geophysical Research Letters, 2012. **39**(9).

393. Anderson, T.L., et al., *Climate forcing by aerosols--a hazy picture.* Science, 2003. **300**(5622): p. 1103-1104.

394. Gottschalk, B., *Global surface temperature trends and the effect of World War II: a parametric analysis (long version).* https://arxiv.org/abs/1703.06511

395. Gottschalk, B., *Global surface temperature trends and the effect of World War II.* https://arxiv.org/abs/1703.9281

396. Stocker, T., et al., *IPCC, 2013: Climate Change 2013: The Physical Science Basis. Contribution of Working Group I to the Fifth Assessment Report of the Intergovernmental Panel on Climate Change, 1535 pp.* 2013, Cambridge Univ. Press, Cambridge, UK, and New York.

397. Archer, D., et al., *Atmospheric lifetime of fossil fuel carbon dioxide.* Annual review of earth and planetary sciences, 2009. **37**: p. 117-134.

398. Bastos, A., et al., *Re-evaluating the 1940s CO2 plateau.* Biogeosciences, 2016. **13**: p. 4877-4897.

399. Müller, J., *Atmospheric residence time of carbonaceous particles and particulate PAH-compounds.* Science of the Total Environment, 1984. **36**: p. 339-346.

400. Poet, S., H. Moore, and E. Martell, *Lead 210, bismuth 210, and polonium 210 in the atmosphere: Accurate ratio measurement and application to aerosol residence time determination.* Journal of Geophysical Research, 1972. **77**(33): p. 6515-6527.

401. Baskaran, M. and G.E. Shaw, *Residence time of arctic haze aerosols using the concentrations and activity ratios of 210Po, 210Pb and 7Be.* Journal of Aerosol Science, 2001. **32**(4): p. 443-452.

402. Quinn, P., et al., *Short-lived pollutants in the Arctic: their climate impact and possible mitigation strategies.* Atmospheric Chemistry and Physics, 2008. **8**(6): p. 1723-1735.

403. Ogren, J. and R. Charlson, *Elemental carbon in the atmosphere: cycle and lifetime.* Tellus B, 1983. **35**(4): p. 241-254.

404. Rutledge, D., *Estimating long-term world coal production with logit and probit transforms.* International Journal of Coal Geology, 2011. **85**(1): p. 23-33.

405. https://www.indexmundi.com/energy/

406. Maggio, G. and G. Cacciola, *When will oil, natural gas, and coal peak?* Fuel, 2012. **98**: p. 111-123.

407. McNeill, J.R., *Something new under the sun: An environmental history of the twentieth-century world (the global century series).* 2001: WW Norton & Company.

408. Carlson, T.N. and S.G. Benjamin, *Radiative heating rates for Saharan dust.* Journal of the Atmospheric Sciences, 1980. **37**(1): p. 193-213.

409. Scortichini, M., et al., *Short-Term Effects of Heat on Mortality and Effect Modification by Air Pollution in 25 Italian Cities.*

International journal of environmental research and public health, 2018. **15**(8): p. 1771.

410. Jacobson, M.Z., *Effects of biomass burning on climate, accounting for heat and moisture fluxes, black and brown carbon, and cloud absorption effects.* Journal of Geophysical Research: Atmospheres, 2014. **119**(14): p. 8980-9002.

411. Ito, A., G. Lin, and J.E. Penner, *Radiative forcing by light-absorbing aerosols of pyrogenetic iron oxides.* Scientific Reports, 2018. **8**(1): p. 7347.

412. Olson, M.R., et al., *Investigation of black and brown carbon multiple-wavelength-dependent light absorption from biomass and fossil fuel combustion source emissions.* Journal of Geophysical Research: Atmospheres, 2015. **120**(13): p. 6682-6697.

413. Oeste, F.D., et al., *Climate engineering by mimicking natural dust climate control: the iron salt aerosol method.* Earth System Dynamics, 2017. **8**(1): p. 1-54.

414. Liu, L., et al., *Cloud scavenging of anthropogenic refractory particles at a mountain site in North China.* Atmospheric Chemistry and Physics, 2018. **18**(19): p. 14681-14693.

415. Hunt, A.J., *Small particle heat exchangers.* University of California, Berkeley Report No. LBL-7841. 1978.

416. Buffett, B.A., *A comparison of subgrid-scale models for large-eddy simulations of convection in the Earth's core.* Geophysical Journal International, 2003. **153**(3): p. 753-765.

417. Roberts, P.H. and E.M. King, *On the genesis of the Earth's magnetism.* Reports on Progress in Physics, 2013. **76**(9): p. 096801.

418. Huguet, L., H. Amit, and T. Alboussière, *Geomagnetic dipole changes and upwelling/downwelling at the top of the Earth's core.* Frontiers in Earth Science, 2018. **6**: p. 170.

419. Glatzmaier, G.A., *Geodynamo simulations - How realistic are they?* Ann. Rev.Earth Planet. Sci., 2002. **30**: p. 237-257.

420. Guervilly, C., P. Cardin, and N. Schaeffer, *Turbulent convective length scale in planetary cores.* Nature, 2019. **570**(7761): p. 368.

421. Tackley, P.J., *Modelling compressible mantle convection with large viscosity contrasts in a three-dimensional spherical shell using the yin-yang grid.* Physics of the Earth and Planetary Interiors, 2008. **171**(1-4): p. 7-18.

422. Gerardi, G., N.M. Ribe, and P.J. Tackley, *Plate bending, energetics of subduction and modeling of mantle convection: A boundary element approach.* Earth and Planetary Science Letters, 2019. **515**: p. 47-57.

423. Nakagawa, T. and H. Iwamori, *On the implications of the coupled evolution of the deep planetary interior and the presence of surface ocean water in hydrous mantle convection.* Comptes Rendus Geoscience, 2019. **351**(2-3): p. 197-208.

424. Herndon, J.M., *Uniqueness of Herndon's georeactor: Energy source and production mechanism for Earth's magnetic field.* https://arxiv.org/abs/0901.4509

425. http://nuclearplanet.com/convection.mp4

426. https://www.youtube.com/watch?v=IAXOXx3H8AI

427. Cao, H.X., J. Mitchell, and J. Lavery, *Simulated diurnal range and variability of surface temperature in a global climate model for present and doubled CO2 climates.* Journal of Climate, 1992. **5**(9): p. 920-943.

428. Qu, M., J. Wan, and X. Hao, *Analysis of diurnal air temperature range change in the continental United States.* Weather and Climate Extremes, 2014. **4**: p. 86-95.

429. Roderick, M.L. and G.D. Farquhar, *The cause of decreased pan evaporation over the past 50 years.* Science, 2002. **298**(5597): p. 1410-1411.

430. Easterling, D.R., et al., *Maximum and minimum temperature trends for the globe.* Science, 1997. **277**(5324): p. 364-367.

431. Dai, A., K.E. Trenberth, and T.R. Karl, *Effects of clouds, soil moisture, precipitation, and water vapor on diurnal temperature range.* Journal of Climate, 1999. **12**(8): p. 2451-2473.

432. Roy, S.S. and R.C. Balling, *Analysis of trends in maximum and minimum temperature, diurnal temperature range, and cloud cover over India.* Geophysical Research Letters, 2005. **32**(12).

433. Peralta-Hernandez, A.R., R.C. Balling Jr, and L.R. Barba-Martinez, *Analysis of near-surface diurnal temperature variations and trends in southern Mexico.* International Journal of Climatology: A Journal of the Royal Meteorological Society, 2009. **29**(2): p. 205-209.

434. Fehler, M. and B. Chouet, *Operation of a digital seismic network on Mount St. Helens volcano and observations of long period seismic events that originate under the volcano.* Geophysical Research Letters, 1982. **9**(9): p. 1017-1020.

435. Mass, C. and A. Robock, *The short-term influence of the Mount St. Helens volcanic eruption on surface temperature in the Northwest United States.* Monthly Weather Review, 1982. **110**(6): p. 614-622.

436. Herndon, J.M., *Science misrepresentation and the climate-science cartel.* J. Geog. Environ. Earth Sci. Intn., 2018. **18**(2): p. 1-13.

437. Talukdar, S., et al., *Influence of black carbon aerosol on the atmospheric instability.* Journal of Geophysical Research: Atmospheres.

438. Dunion, J.P. and C.S. Velden, *The impact of the Saharan air layer on Atlantic tropical cyclone activity.* Bulletin of the American Meteorological Society, 2004. **85**(3): p. 353-366.

439. Prospero, J.M. and T.N. Carlson, *Vertical and areal distribution of Saharan dust over the western equatorial North Atlantic Ocean.* Journal of Geophysical Research, 1972. **77**(27): p. 5255-5265.

440. Alfaro, S., et al., *Iron oxides and light absorption by pure desert dust: An experimental study.* Journal of Geophysical Research: Atmospheres, 2004. **109**(D8).

441. Liu, D., et al., *Aircraft and ground measurements of dust aerosols over the west African coast in summer 2015 during ICE-D and AER-D.* Atmospheric Chemistry and Physics, 2018. **18**(5): p. 3817-3838.

442. Walsh, J.J. and K.A. Steidinger, *Saharan dust and Florida red tides: the cyanophyte connection.* Journal of Geophysical Research: Oceans, 2001. **106**(C6): p. 11597-11612.

443. Wang, R., et al., *Sources, transport and deposition of iron in the global atmosphere.* Atmospheric Chemistry and Physics, 2015. **15**(11): p. 6247-6270.

444. Wells, M., L. Mayer, and R. Guillard, *Evaluation of iron as a triggering factor for red tide blooms.* Marine ecology progress series, 1991: p. 93-102.

445. Wong, S. and A.E. Dessler, *Suppression of deep convection over the tropical North Atlantic by the Saharan Air Layer.* Geophysical research letters, 2005. **32**(9).

446. Landsberg, H.E., *The Urban Climate, Volume 28.* 1981, Academic Press.

447. Roth, M., T. Oke, and W. Emery, *Satellite-derived urban heat islands from three coastal cities and the utilization of such data in urban climatology.* International Journal of Remote Sensing, 1989. **10**(11): p. 1699-1720.

448. Hua, L., Z. Ma, and W. Guo, *The impact of urbanization on air temperature across China.* Theoretical and Applied Climatology, 2008. **93**(3-4): p. 179-194.

449. Alcoforado, M.J. and H. Andrade, *Global warming and the urban heat island*, in *Urban ecology*. 2008, Springer. p. 249-262.

450. Herndon, J.M. and M. Whiteside, *Geoengineering: The deadly new global "Miasma".* Journal of Advances in Medicine and Medical Research, 2019. **29**(12): p. 1-8.

451. Fan, J., et al., *Review of aerosol–cloud interactions: Mechanisms, significance, and challenges.* Journal of the Atmospheric Sciences, 2016. **73**(11): p. 4221-4252.

452. Pöschl, U., *Atmospheric aerosols: composition, transformation, climate and health effects.* Angewandte Chemie International Edition, 2005. **44**(46): p. 7520-7540.

453. Ito, A., *Atmospheric processing of combustion aerosols as a source of bioavailable iron.* Environmental Science & Technology Letters, 2015. **2**(3): p. 70-75.

454. Ito, A., et al., *Pyrogenic iron: The missing link to high iron solubility in aerosols.* Science Advances, 2019. **5**(5): p. eaau7671.

455. Matsui, H., et al., *Anthropogenic combustion iron as a complex climate forcer.* Nature communications, 2018. **9**(1): p. 1593.

456. Moteki, N., et al., *Anthropogenic iron oxide aerosols enhance atmospheric heating.* Nature communications, 2017. **8**: p. 15329.

457. Latham, J., et al., *Marine cloud brightening.* Philosophical Transactions of the Royal Society A: Mathematical, Physical and Engineering Sciences, 2012. **370**(1974): p. 4217-4262.

458. Kumar, M., et al., *Carbon dioxide capture, storage and production of biofuel and biomaterials by bacteria: A review.* Bioresource technology, 2018. **247**: p. 1059-1068.

459. Reynolds, J.L., *Solar geoengineering to reduce climate change: a review of governance proposals.* Proceedings of the Royal Society A, 2019. **475**(2229): p. 20190255.

460. World Health Organization, *Ambient air pollution – a global assessment of exposure and burden of disease.*, in *WHO Library Cataloguing-in-Publication.* 2016, Geneva.

461. Apte, J.S., et al., *Ambient PM2.5 Reduces Global and Regional Life Expectancy.* Environmental Science & Technology Letters, 2018.

462. Pope, A., et al., *Lung cancer, cardiopulmonary mortality, and long-term exposure to fine particulate air pollution.* JAMA, 2002. **287**(9): p. 1132-1141.

463. Genc, S., et al., *The Adverse Effects of Air Pollution on the Nervous System.* Journal of Toxicology, 2012. **2012**.

464. Calderón-Garcidueñas, L., et al., *Air pollution, cognitive deficits and brain abnormalities: a pilot study with children and dogs.* Brain and cognition, 2008. **68**(2): p. 117-127.

465. Whiteside, M. and J.M. Herndon, *Humic like substances (HULIS): Contribution to global warming and stratospheric ozone depletion.* European Journal of Applied Sciences, 2023. **11**(2): p. 325-346.

466. Whiteside, M. and J.M. Herndon, *Disruption of Earth's atmospheric flywheel: Hothouse-Earth collapse of the biosphere and causation of the sixth great extinction.* European Journal of Applied Sciences, 2024. **12**(1): p. 361-395.

467. Fabry, C., *Ozone as an Absorbing Material for Radiations in the Atmosphere.* Journal of Mathematics and Physics, 1925. **4**(1-4): p. 1-20.

468. Farman, J.C., B.G. Gardiner, and J.D. Shanklin, *Large losses of total ozone in Antarctica reveal seasonal ClOx/NOx interaction.* Nature, 1985. **315**(6016): p. 207-210.

469. Stolarski, R.S., et al., *Nimbus 7 satellite measurements of the springtime Antarctic ozone decrease.* Nature, 1986. **322**(6082): p. 808-811.

470. Hamill, P., O. Toon, and R. Turco, *Characteristics of polar stratospheric clouds during the formation of the Antarctic ozone hole.* Geophysical research letters, 1986. **13**(12): p. 1288-1291.

471. Solomon, S., *The mystery of the Antarctic ozone "hole".* Reviews of Geophysics, 1988. **26**(1): p. 131-148.

472. Molina, M.J. and F.S. Rowland, *Stratospheric sink for chlorofluoromethanes: chlorine atom-catalysed destruction of ozone.* Nature, 1974. **249**(5460): p. 810-812.

473. Crutzen, P.J., *Ozone production rates in an oxygen-hydrogen-nitrogen oxide atmosphere.* Journal of Geophysical Research, 1971. **76**(30): p. 7311-7327.

474. Tritscher, I., et al., *Polar stratospheric clouds: Satellite observations, processes, and role in ozone depletion.* Reviews of geophysics, 2021. **59**(2): p. e2020RG000702.

475. Solomon, S., et al., *On the depletion of Antarctic ozone.* Nature, 1986. **321**(6072): p. 755-758.

476. https://www.unep.org/ozonaction/who-we-are/about-montreal-protocol

477. Córdoba, C., et al., *The detection of solar ultraviolet-C radiation using KCl:Eu2+ thermoluminescence dosemeters.* Journal of Physics D: Applied Physics, 1997. **30**(21): p. 3024.

478. D'Antoni, H., et al., *Extreme environments in the forests of Ushuaia, Argentina.* Geophysical Research Letters, 2007. **34**(22).

479. Witze, A., *Rare ozone hole opens over Arctic--and it's big.* Nature, 2020. **580**(7801): p. 18-20.

480. Lu, Q.-B., *Observation of large and all-season ozone losses over the tropics.* AIP Advances, 2022. **12**(7): p. 075006.

481. Bernhard, G.H., et al., *Updated analysis of data from Palmer Station, Antarctica (64° S), and San Diego, California (32° N), confirms large effect of the Antarctic ozone hole on UV radiation.* Photochemical & Photobiological Sciences, 2022. **21**(3): p. 373-384.

482. Cordero, R.R., et al., *Persistent extreme ultraviolet irradiance in Antarctica despite the ozone recovery onset.* Scientific reports, 2022. **12**(1): p. 1-10.

483. Takahashi, T., et al., *Measurement of solar UV radiation in antarctica with collagen sheets.* Photochemical & Photobiological Sciences, 2012. **11**(7): p. 1193-1200.

484. Wignall, P.B., *Large igneous provinces and mass extinctions.* Earth-Science Reviews, 2001. **53**(1): p. 1-33.

485. Ogden, D.E. and N.H. Sleep, *Explosive eruption of coal and basalt and the end-Permian mass extinction.* Proceedings of the National Academy of Sciences, 2012. **109**(1): p. 59-62.

486. Grasby, S.E., H. Sanei, and B. Beauchamp, *Catastrophic dispersion of coal fly ash into oceans during the latest Permian extinction.* Nature Geoscience, 2011. **4**(2): p. 104.

487. Brand, U., et al., *Methane Hydrate: Killer cause of Earth's greatest mass extinction.* Palaeoworld, 2016. **25**(4): p. 496-507.

488. Visscher, H., et al., *Environmental mutagenesis during the end-Permian ecological crisis.* Proceedings of the National Academy of Sciences of the United States of America, 2004. **101**(35): p. 12952-12956.

489. Schoene, B., et al., *U-Pb geochronology of the Deccan Traps and relation to the end-Cretaceous mass extinction.* Science, 2015. **347**(6218): p. 182-184.

490. Huang, S.-H. and C.-C. Chen, *Ultrafine aerosol penetration through electrostatic precipitators.* Environmental science & technology, 2002. **36**(21): p. 4625-4632.

491. Baxter, M., *Environmental radioactivity: A perspective on industrial contributions.* IAEA Bulletin, 1993. **35**(2): p. 33-38.

492. Umo, N.S., et al., *Enhanced ice nucleation activity of coal fly ash aerosol particles initiated by ice-filled pores.* Atmospheric chemistry and physics, 2019. **19**(13): p. 8783-8800.

493. Cziczo, D.J., et al., *Clarifying the dominant sources and mechanisms of cirrus cloud formation.* Science, 2013. **340**(6138): p. 1320-1324.

494. Richardson, M.S., et al., *Measurements of heterogeneous ice nuclei in the western United States in springtime and their relation to aerosol characteristics.* Journal of Geophysical Research: Atmospheres, 2007. **112**(D2).

495. Das, T., B.K. Saikia, and B.P. Baruah, *Formation of carbon nano-balls and carbon nano-tubes from northeast Indian Tertiary coal: value added products from low grade coal.* Gondwana Research, 2016. **31**: p. 295-304.

496. Alam, J., et al., *Recent advances in methods for the recovery of carbon nanominerals and polyaromatic hydrocarbons from coal fly ash and their emerging applications.* Crystals, 2021. **11**(2): p. 88.

497. Francis, A.H., *Electronic Structure Calculations on Fullerenes and Their Derivatives By Jerzy Cioslowski (Florida State University). Oxford University Press: New York. 1995. ix + 281 pp. $65.00. ISBN 0-19-508806-9.* Journal of the American Chemical Society, 1996. **118**(39): p. 9458-9458.

498. Dosodia, A., et al., *Development of Catalyst Free Carbon Nanotubes from Coal and Waste Plastics.* Fullerenes, Nanotubes and Carbon Nanostructures, 2009. **17**(5): p. 567-582.

499. Tiwari, A.J., M. Ashraf-Khorassani, and L.C. Marr, *C60 fullerenes from combustion of common fuels.* Science of The Total Environment, 2016. **547**: p. 254-260.

500. Saikia, J., et al., *Polycyclic aromatic hydrocarbons (PAHs) around tea processing industries using high-sulfur coals.* Environmental Geochemistry and Health, 2017. **39**(5): p. 1101-1116.

501. Hower, J.C., et al., *Association of the Sites of Heavy Metals with Nanoscale Carbon in a Kentucky Electrostatic Precipitator Fly Ash.* Environmental Science & Technology, 2008. **42**(22): p. 8471-8477.

502. Paul, K.T., et al., *Preparation and Characterization of Nano structured Materials from Fly Ash: A Waste from Thermal Power Stations, by High Energy Ball Milling.* Nanoscale Research Letters, 2007. **2**(8): p. 397.

503. Graham, U., et al. *Ultra-Fine PM Derived from Fullerene-Like Carbon in Electrostatic Precipitator Fly Ash.* in *Proceedings of 2008 AIChE Annual Meeting, Philadelphia (USA).* 2008.

504. Salah, N., et al., *Formation of Carbon Nanotubes from Carbon-Rich Fly Ash: Growth Parameters and Mechanism.* Materials and Manufacturing Processes, 2016. **31**(2): p. 146-156.

505. Monthioux, M. and V.L. Kuznetsov, *Who should be given the credit for the discovery of carbon nanotubes?* Carbon, 2006. **44**(9): p. 1621-1623.

506. Kronbauer, M.A., et al., *Geochemistry of ultra-fine and nano-compounds in coal gasification ashes: A synoptic view.* Science of The Total Environment, 2013. **456-457**: p. 95-103.

507. Chen, Y., et al., *Transmission electron microscopy investigation of ultrafine coal fly ash particles.* Environ. Science and Technogy, 2005. **39**(4): p. 1144-1151.

508. Murr, L.E. and K.F. Soto, *A TEM study of soot, carbon nanotubes, and related fullerene nanopolyhedra in common fuel-gas combustion sources.* Materials Characterization, 2005. **55**(1): p. 50-65.

509. Moon, M.-W., et al., *Nanostructured Carbon Materials.* Journal of Nanomaterials, 2015. **2015**: p. 916834.

510. Everson, R.C., et al., *Reaction kinetics of pulverized coal-chars derived from inertinite-rich coal discards: Gasification with carbon dioxide and steam.* Fuel, 2006. **85**(7): p. 1076-1082.

511. Chen, Z., et al., *Energy Storage: Confined Assembly of Hollow Carbon Spheres in Carbonaceous Nanotube: A Spheres-in-Tube Carbon Nanostructure with Hierarchical Porosity for High-*

Performance Supercapacitor (Small 19/2018). Small, 2018. **14**(19): p. 1870089.

512. Silva, L., T. Moreno, and X. Querol, *An introductory TEM study of Fe-nanominerals within coal fly ash.* Science of the Total Environment, 2009. **407**(17): p. 4972-4974.

513. Chen, Y., et al., *Characterization of ultrafine coal fly ash particles by energy filtered TEM.* Journal of Microscopy, 2005. **217**(3): p. 225-234.

514. Martinello, K., et al., *Direct identification of hazardous elements in ultra-fine and nanominerals from coal fly ash produced during diesel co-firing.* Science of the Total Environment, 2014. **470**: p. 444-452.

515. Ribeiro, J., et al., *Extensive FE-SEM/EDS, HR-TEM/EDS and ToF-SIMS studies of micron-to nano-particles in anthracite fly ash.* Science of the total environment, 2013. **452**: p. 98-107.

516. Silva, L.F., et al., *Fullerenes and metallofullerenes in coal-fired stoker fly ash.* Coal Combustion and Gasification Products, 2010. **2**: p. 66-79.

517. Oliveira, M.L., et al., *Nano-mineralogical investigation of coal and fly ashes from coal-based captive power plant (India): an introduction of occupational health hazards.* Science of the Total Environment, 2014. **468**: p. 1128-1137.

518. Dias, C.L., et al., *Nanominerals and ultrafine particles from coal fires from Santa Catarina, South Brazil.* International Journal of Coal Geology, 2014. **122**: p. 50-60.

519. Linak, W.P., et al., *Ultrafine ash aerosols from coal combustion: Characterization and health effects.* Proceedings of the Combustion Institute, 2007. **31**(2): p. 1929-1937.

520. Schütze, K., et al., *Submicrometer refractory carbonaceous particles in the polar stratosphere.* 2017.

521. de Reus, M., et al., *Particle production in the lowermost stratosphere by convective lifting of the tropopause.* Journal of Geophysical Research: Atmospheres, 1999. **104**(D19): p. 23935-23940.

522. Baars, H., et al., *The unprecedented 2017–2018 stratospheric smoke event: decay phase and aerosol properties observed with the EARLINET.* Atmospheric chemistry and physics, 2019. **19**(23): p. 15183-15198.

523. Nielsen, J.K., et al., *Solid particles in the tropical lowest stratosphere.* Atmospheric Chemistry and Physics, 2007. **7**(3): p. 685-695.

524. Ebert, M., et al., *Chemical analysis of refractory stratospheric aerosol particles collected within the arctic vortex and inside polar stratospheric clouds.* Atmospheric Chemistry and Physics, 2016. **16**(13): p. 8405-8421.

525. Simpson, W.R., et al., *Halogens and their role in polar boundary-layer ozone depletion.* Atmospheric Chemistry and Physics, 2007. **7**(16): p. 4375-4418.

526. Read, K.A., et al., *Extensive halogen-mediated ozone destruction over the tropical Atlantic Ocean.* Nature, 2008. **453**(7199): p. 1232-1235.

527. NRC, *Trace-element Geochemistry of Coal Resource Development Related to Environmental Quality and Health.* 1980: National Academy Press.

528. Peng, X., et al., *An unexpected large continental source of reactive bromine and chlorine with significant impact on wintertime air quality.* National science review, 2021. **8**(7): p. nwaa304.

529. Pedersen, K.H., et al., *Post-treatment of fly ash by ozone in a fixed bed reactor.* Energy & fuels, 2009. **23**(1): p. 280-285.

530. Chen, X., et al. *Fly ash beneficiation with ozone: Mechanism of absorption suppression.* in *Abstracts of papers of the American*

Chemical Society. 2002. Amer Chemical Soc 1155 16TH ST, NW, Washington, DC 20036 USA.

531. Alebic-Juretic, A., T. Cvitas, and L. Klasinc, *Ozone destruction on solid particles.* Environmental monitoring and assessment, 1997. **44**(1): p. 241-247.

532. Atale, S., et al., *Ozone reactions with various carbon materials.* Jap Pat CA, 1995. **123**: p. 121871.

533. Zhang, H., J.Y. Lee, and H. Liu, *Ozone Decomposition on Defective Graphene: Insights from Modeling.* The Journal of Physical Chemistry C, 2021. **125**(20): p. 10948-10954.

534. Michel, A., C. Usher, and V. Grassian, *Reactive uptake of ozone on mineral oxides and mineral dusts.* Atmospheric Environment, 2003. **37**(23): p. 3201-3211.

535. Coates Fuentes, Z.L., T.M. Kucinski, and R.Z. Hinrichs, *Ozone decomposition on kaolinite as a function of monoterpene exposure and relative humidity.* ACS Earth and Space Chemistry, 2018. **2**(1): p. 21-30.

536. Lasne, J., M.N. Romanias, and F. Thevenet, *Ozone uptake by clay dusts under environmental conditions.* ACS Earth and Space Chemistry, 2018. **2**(9): p. 904-914.

537. Hanisch, F. and J. Crowley, *Ozone decomposition on Saharan dust: an experimental investigation.* Atmospheric Chemistry and Physics Discussions, 2002. **2**(6): p. 1809-1845.

538. Yan, L., J. Bing, and H. Wu, *The behavior of ozone on different iron oxides surface sites in water.* Scientific reports, 2019. **9**(1): p. 1-10.

539. Xu, Z., et al., *A novel γ-like MnO2 catalyst for ozone decomposition in high humidity conditions.* Journal of Hazardous Materials, 2021. **420**: p. 126641.

540. Heisig, C., W. Zhang, and S.T. Oyama, *Decomposition of ozone using carbon-supported metal oxide catalysts.* Applied catalysis B: environmental, 1997. **14**(1-2): p. 117-129.

541. Kashtanov, L., N. Ivanova, and B. Rizhov, *Catalytic activity of metals in ozone decomposition.* J. Applied Chemistry, 1936. **9**: p. 2176-2182.

542. Reckhow, D.A., et al., *Oxidation Of Iron And Manganese By Ozone.* Ozone: Science & Engineering, 1991. **13**(6): p. 675-695.

543. Emelyanova, G., V. Lebedev, and N. Kobozev, *Catalytic activity of noble metals in ozone destruction.* J Phys Chem, 1964. **38**: p. 170-180.

544. Herndon, J.M., *Evidence of coal-fly-ash toxic chemical geoengineering in the troposphere: Consequences for public health* Int. J. Environ. Res. Public Health 2015. **12**(8).

545. Herndon, J.M. and M. Whiteside, *Further evidence of coal fly ash utilization in tropospheric geoengineering: Implications on human and environmental health.* J. Geog. Environ. Earth Sci. Intn., 2017. **9**(1): p. 1-8.

546. Herndon, J.M. and M. Whiteside, *Further evidence of coal fly ash utilization in tropospheric geoengineering: Implications on human and environmental health.* J. Geog. Environ. Earth Sci. Intn., 2017. **9**(1): p. 1-8.

547. Whiteside, M. and J.M. Herndon, *Aerosol particulates, SARS-Co-2, and the broader potential for global devastation.* Open Access Journal of Internal Medicine, 2022. **3**(1): p. 14-21.

548. Ball, W.T., et al., *Evidence for a continuous decline in lower stratospheric ozone offsetting ozone layer recovery.* Atmospheric Chemistry and Physics, 2018. **18**(2): p. 1379-1394.

549. Cohen, W., *Address by Defense Secretary Cohen: Terrorism, Weapons of Mass Destruction, and U. S. Strategy*, in *Sam Nunn Policy Forum, University of Georgia, Athens, Georgia, April 28, 1997*. 1997.

550. *https://irp.fas.org/program/collect/haarp-duma.htm*

551. Schmitter, E.D., *Signals from the Rocks*. International Journal of Geotechnical and Geological Engineering, 2008. **2**(1): p. 1-4.

552. Nardi, A., M. Caputo, and C. Chiarabba, *Possible electromagnetic earthquake precursors in two years of ELF-VLF monitoring in the atmosphere*. Boll. Geofis. Teor. Appl, 2007. **48**(2): p. 205-212.

553. O'keefe, S. and D. Thiel, *Electromagnetic emissions during rock blasting*. Geophysical Research Letters, 1991. **18**(5): p. 889-892.

554. Frid, V. and K. Vozoff, *Electromagnetic radiation induced by mining rock failure*. International Journal of Coal Geology, 2005. **64**(1-2): p. 57-65.

555. Zhao, S., et al., *A lithosphere-atmosphere-ionosphere coupling model for ELF electromagnetic waves radiated from seismic sources and its possibility observed by the CSES*. Science China Technological Sciences, 2021. **64**(11): p. 2551-2559.

556. Gheonjian, L., T. Paatashvili, and G. Kapanadze. *ELF radio emission associated with strong M6.0 earthquake*. in *2017 XXIInd International Seminar/Workshop on Direct and Inverse Problems of Electromagnetic and Acoustic Wave Theory (DIPED)*. 2017.

557. Serebryakova, O., et al., *Electromagnetic ELF radiation from earthquake regions as observed by low-altitude satellites*. Geophysical Research Letters, 1992. **19**(2): p. 91-94.

558. Christofilakis, V., et al., *Significant ELF perturbations in the Schumann Resonance band before and during a shallow mid-magnitude seismic activity in the Greek area (Kalpaki)*. Journal of Atmospheric and Solar-Terrestrial Physics, 2019. **182**: p. 138-146.

559. Chmyrev, V., et al., *Small-scale plasma inhomogeneities and correlated ELF emissions in the ionosphere over an earthquake region*. Journal of Atmospheric and Solar-Terrestrial Physics, 1997. **59**(9): p. 967-974.

560. Wang, Y., X. Yuan, and Q. Cao. *Study of earthquake location using electromagnetic precursors*. in *2013 Asia-Pacific Microwave Conference Proceedings (APMC)*. 2013. IEEE.

561. Bhattacharya, S., et al., *Satellite and ground-based ULF/ELF emissions observed before Gujarat earthquake in March 2006.* Current Science, 2007: p. 41-46.

562. *https://www.youtube.com/watch?app=desktop&v=PN8cHwamcJA*

563. Jiang, C., et al., *An overview of resonant circuits for wireless power transfer.* Energies, 2017. **10**(7): p. 894.

564. Greifinger, P., V. Mushtak, and E. Williams. *The lower characteristic ELF altitude of the Earth-ionosphere waveguide: Schumann resonance observations and aeronomical estimates.* in *IEEE 6th International Symposium on Electromagnetic Compatibility and Electromagnetic Ecology, 2005.* 2005. IEEE.

565. Cohen, M. and M. Gołkowski, *100 days of ELF/VLF generation via HF heating with HAARP.* Journal of Geophysical Research: Space Physics, 2013. **118**(10): p. 6597-6607.

566. Papadopoulos, K., et al., *First demonstration of HF-driven ionospheric currents.* Geophysical research letters, 2011. **38**(20).

567. Barr, R., et al., *ELF and VLF signals radiated by the "polar electrojet antenna": experimental results.* Journal of Geophysical Research: Space Physics, 1986. **91**(A4): p. 4451-4459.

568. McCarrick, M., et al., *Excitation of ELF waves in the Schumann resonance range by modulated HF heating of the polar electrojet.* Radio science, 1990. **25**(6): p. 1291-1298.

569. Stubbe, P., et al., *Ionospheric modification experiments with the Tromsø heating facility.* Journal of atmospheric and terrestrial physics, 1985. **47**(12): p. 1151-1163.

570. Ferraro, A. and D. Werner. *A technique using an ionospheric modification instrument to produce controllable and steerable low frequency ionospheric arrays for remote sensing applications.* in *1995 Ninth International Conference on Antennas and Propagation, ICAP'95 (Conf. Publ. No. 407).* 1995. IET.

571. Cohen, M.B., et al., *Magnetospheric injection of ELF/VLF waves with modulated or steered HF heating of the lower ionosphere.* Journal of Geophysical Research: Space Physics, 2011. **116**(A6).

572. Guo, Z., H. Fang, and F. Honary, *The generation of ULF/ELF/VLF waves in the ionosphere by modulated heating.* Universe, 2021. **7**(2): p. 29.

573. Lunnen, R., et al., *Detection of local and long-path VLF/ELF radiation from modulated ionospheric current systems.* Radio Science, 1985. **20**(3): p. 553-563.

574. Gołkowski, M., et al., *Magnetospheric amplification and emission triggering by ELF/VLF waves injected by the 3.6 MW HAARP ionospheric heater.* Journal of Geophysical Research: Space Physics, 2008. **113**(A10).

575. Yampolski, Y., et al., *Ionospheric non-linear effects observed during very-long-distance HF propagation.* Frontiers in Astronomy and Space Sciences, 2019. **6**: p. 12.

576. Tarasov, N., et al., *The effect of high energy electromagnetic pulses on seismicity in Central Asia and Kazakhstan.* 2000.

577. Avagimov, A. and V. Zeigarnik, *The analysis of the trigger action exerted by electromagnetic fields on a geological medium: Quantitative estimates of the interaction.* Izvestiya, Physics of the Solid Earth, 2016. **52**: p. 233-241.

578. De Aquino, F., *High-power ELF radiation generated by modulated HF heating of the ionosphere can cause Earthquakes, Cyclones and localized heating.* HAL (Open Science) 2011 hal-01082992 2011.

579. *https://centerforinquiry.org/video/debunking-9-11-conspiracy-theories-mick-west/*

580. *https://www.metabunk.org/home/*

581. *https://contrailscience.com/*

582. *http://nuclearplanet.com/Public_Deception_by_Scientists.html*

583. https://www.metabunk.org/threads/debunked-wtc-towers-fell-in-their-own-footprints.1226/

584. https://twitter.com/MickWest/status/1031713999203647488

585. https://www.facebook.com/ae911truth/posts/mick-west-owner-of-the-website-metabunk-is-perhaps-the-most-influential-demoliti/10157118390106269/

586. https://www.metabunk.org/threads/evergreen-air-geoengineeringwatch-org.653/

587. https://groups.google.com/g/geoengineering/c/A0_htKZgLos/m/xWp1gGdlBH4J

588. https://www.logically.ai/factchecks/library/6b82df13

589. https://www.metabunk.org/threads/haarp-returns-holds-open-house.7867/

590. Golkowski, M., U.S. Inan, and M.B. Cohen, *Cross modulation of whistler mode and HF waves above the HAARP ionospheric heater.* Geophysical research letters, 2009. **36**(15).

591. Rietveld, M.T. and P. Stubbe, *History of the Tromsø ionosphere heating facility.* Hist. Geo Space. Sci., 2022. **13**(1): p. 71-82.

592. https://www.youtube.com/watch?v=SZuxt8yZ5Yc

593. https://www.youtube.com/watch?v=Y9DflVY0mHk

594. https://icecube.wisc.edu/science/icecube/#:~:text=IceCube%2C%20the%20only%20cubic%2Dkilometer,for%20the%20scientific%20research%20program

595. https://wikileaks.org/clinton-emails/emailid/11791

596. *https://nzhistory.govt.nz/media/photo/christchurch-earthquake-images*

597. *https://knowledge.aidr.org.au/resources/earthquake-christchurch-new-zealand-2011/*

598. *https://www.pjstar.com/story/news/2011/02/22/u-s-sends-rescue-team/42446617007/*

599. *https://www.usgs.gov/programs/earthquake-hazards/news/new-interactive-geonarrative-explains-2023-turkey-earthquake*

600. *https://www.trthaber.com/haber/gundem/asrin-felaketinde-can-kaybi-47-bin-932-oldu-752123.html*

601. *https://reliefweb.int/report/syrian-arab-republic/syriaturkey-earthquakes-situation-report-7-march-8-2023*

602. *https://www.reuters.com/world/middle-east/earthquake-death-toll-surpasses-50000-turkey-syria-2023-02-24/*

603. *https://www.ungeneva.org/en/news-media/news/2023/02/78128/15-million-now-homeless-turkiye-after-quake-disaster-warn-un*

604. *https://www.aljazeera.com/news/2023/2/3/turkey-says-western-nations-gave-no-evidence-of-security-risks*

605. *https://www.voltairenet.org/article218842.html*

606. *https://www.bitchute.com/video/XhI8JovX9UkZ/*

607. *https://www.usatoday.com/story/news/factcheck/2023/02/15/fact-check-false-claim-haarp-responsible-turkey-earthquakes/11220907002/*

608. *https://en.dailypakistan.com.pk/07-Feb-2023/haarp-conspiracy-theories-link-controversial-us-weapon-to-earthquake-in-turkiye-syria*

609. *https://www.metabunk.org/threads/debunked-haarp-elf-waves-causing-a-earthquake.5866/*

610. *https://www.reuters.com/news/archive/factCheckNew*

611. *https://www.politifact.com/factchecks/2023/feb/15/facebook-posts/turkey-syria-earthquakes-were-natural-disaster-not/*

612. *https://www.voltairenet.org/article218842.html#nb2*

613. *https://www.voltairenet.org/article218844.html*

614. Herndon, J.M., M. Whiteside, and I. Baldwin, *Open letter to the International Criminal Court alleging war crimes and crimes against humanity for ntentional triggering the February 6, 2023 earthquake in the Republic of Türkiye.* Advances in Social Science Research, 2023. **10**(3): p. 168-179.

615. *https://seymourhersh.substack.com/p/how-america-took-out-the-nord-stream*

INDEX

Abee, 21, 22, 23, 24, 61, 173
ad hoc assumption, 9, 27, 36, 58, 75, 99
admirable speculation, 111
adverse temperature gradient, 76, 139, 140, 141, 143, 144
aerosol particles, 133, 141, 143, 153, 155, 211, 215
agriculture, 65, 131, 166, 184, 199
Ahmadinejad, 128, 198
air pollution, 119, 123, 126, 147, 197, 208, 209
Aljazeera, 163
aluminum, 67, 119, 120, 196
American Geophysical Union, 80, 174
American Physical Society, 33
Anderson, 27, 175, 186, 189, 202
Andreae, 136, 201, 202
Antarctic Law Dome, 137
anthropogenic, 36, 49, 135, 136, 145, 147, 152, 166, 195, 199, 202, 204
antineutrino, 28, 31, 71, 176, 177
Arctic ice, 132

Aristarchus, 7
arsenic, 125
arsenopyrite, 52
asteroid belt, 65, 81, 84, 85
asteroid impact, 35, 77
Astrophysical Journal Letters, 59, 109, 110
astrophysics, iii, 104, 106, 110
Atlantic, 144, 206, 207, 215
atmospheric, 26, 70, 91, 102, 119, 126, 129, 133, 135, 139, 143, 144, 146, 153, 194, 195, 200, 201, 202, 206, 208, 209, 219
atmospheric aerosols, 201
atmospheric convection, 139, 143, 144, 146, 200
banded ironstone, 40
barium, 100, 119, 120, 155
basalt flood, 48, 51, 69, 72, 73
basaltic, 26, 67, 152, 176, 178
bats, 127, 156
Becquerel, 98
Bethe, 99, 100, 104, 191
big bang, 112
Bill of Rights, 166

biota, 10, 36, 124, 146, 196
Birch, 16, 17, 20, 173, 178
birds, 115, 127, 156
black carbon, 135, 136, 144, 201, 202, 206
black holes, 106, 107
blacklisted, 110
Bohr, 2
Box, 36, 170
breeder reactor, 25, 73, 75
British Commonwealth, iii, 119, 123
bromine, 153, 215
Bulletin of the World Health Organization, 145
Burbidge, 111, 193
Burnett, 24, 174
calcium, 17, 22, 23, 54, 67
calcium sulfide, 22, 23
California, 1, 46, 75, 80, 123, 126, 192, 195, 204, 210
California Institute of Technology, 80
carbon dioxide, 133, 134, 135, 138, 145, 147, 152, 202, 213
carbonaceous chondrite, 60, 68, 82, 83
carcinogens, 125
cardiovascular disease, 127, 147, 156
cartel, 13, 34, 37, 200, 206
cellular redox balance, 125
Chamberlain, 57
Chandrasekhar, 139, 140, 189
chemically mobile aluminum, 125, 126
chemtrails, 118, 119, 120, 125, 127, 130, 159, 194, 199
chlorine, 149, 153, 210, 215
chlorofluorocarbons, 150, 154
Christchurch, New Zealand, 162, 165
climate change, 133, 136, 142, 145, 199, 202, 208
Clinton, 162, 165
clouds, 101, 117, 135, 141, 142, 148, 153, 155, 156, 198, 206, 209, 210, 215
CO2, 133, 134, 137, 202
coal burning, 120, 152
coal fly ash, 120, 121, 122, 123, 124, 125, 126, 127, 129, 130, 132, 145, 146, 148, 151, 152, 153, 154, 155, 156, 195, 196, 197, 198, 211, 212, 213, 214, 217
composition of the sun, 17, 57
condensate, 54, 57, 58, 61, 82, 86, 91
condense, 17, 59, 81
consensus, 3, 13, 14, 142, 172, 194
continent, 9, 37, 46, 52
continent splitting, 52
contrails, 118, 119, 155, 195
convection, 11, 32, 33, 34, 35, 71, 74, 75, 76, 77, 78, 79, 139, 140, 141, 143, 144, 181, 202, 204, 205, 207
COPD, 156, 197
Copernicus, 7, 170
corals, 127
core, 11, 14, 16, 17, 18, 20, 22, 23, 24, 25, 26, 29, 31, 32, 33, 34, 39, 47, 53, 54, 61, 62, 63, 64, 71, 72, 74, 76, 81, 86, 91, 137, 140, 174, 175, 204
corona mass ejection, 35
crimes against humanity, 163, 223
crust, 14, 38, 39, 40, 41, 47, 49, 50, 52, 54, 74, 79, 94, 96, 158

cryoconite, 132
Curie temperature, 31
Current Science, 26, 31, 120, 219
Curry, 134, 201
cyanobacterial blooms, 127
dark galaxy, 107, 108
Deccan Traps, 48, 69, 72, 187, 211
decompression, 40, 41, 44, 45, 47, 49, 51, 52, 54, 55, 79, 86, 89, 92, 93, 94, 95, 96, 167, 169, 178, 187
decompression cracks, 41, 44, 49, 51, 54, 86, 89, 93
Deep State, i, iii, 14, 34, 36, 55, 97, 113, 115, 157, 159, 167
dementia, 35
desiccation, 126
Die Naturwissenschaften, 100, 192
diurnal temperature range, 142, 143, 206
Doppler shift, 111
Doppler shifts, 111
droughts, 128, 157
DTR, 142, 143
dynamo, 11, 32, 33, 74, 75, 190
Earth expansion theory, 37
Earth-facing side, 73
earthquake, 14, 15, 16, 74, 159, 162, 163, 164, 165, 173, 218, 219, 222, 223
East African Rift System, 49, 50, 51, 93
ecosystems, 115, 150
Einstein, 11, 171
electromagnetic radiation, 112, 129, 158, 159
Elsasser, 11, 32, 33, 76, 171
endo-Earth, 20, 23
energy, 25, 28, 29, 30, 31, 37, 38, 39, 41, 47, 52, 53, 54, 69, 71, 75, 76, 77, 79, 80, 95, 96, 97, 98, 99, 100, 102, 112, 113, 134, 136, 138, 139, 145, 157, 158, 163, 168, 176, 193, 203, 214, 220
ENMOD, 130, 131, 132, 133, 145, 146, 199
enstatite, 17, 20, 21, 22, 23, 24, 54, 60, 61, 62, 68, 82, 83, 173, 174
enstatite chondrite, 17, 20, 21, 22, 23, 24, 60, 61, 62, 68, 82, 83, 173
environmental health, 10, 118, 125, 129, 130, 147, 148, 156, 197, 199, 217
environmental instabilities, 157
environmental modification, 121, 131, 145, 157, 166
environmental warfare, 128, 130, 133
Eos, Transactions, American Geophysical Union, 80
Eric Hecker, 161, 162
ethics, 2, 80
Eucken, 60, 61, 81, 184
European Union, iii, 119, 123
exoplanets, 59
Faraday, 11, 32, 77, 171
Fermi, 25, 72, 101, 102, 174
fictitious super-continent cycles, 55
fission, 25, 33, 41, 47, 49, 69, 76, 100, 101, 102, 103, 104, 168
fjords, 44, 45, 55, 69, 89, 90, 96
floods, 72, 79, 157, 187
fluid core, 11, 15, 32, 34, 38, 48, 54, 70, 74, 75, 139, 169
fluorine, 149, 153
forensic science, 119

forest die-offs, 196, 198
forest fires, 126, 130
Frank, 39, 178
frost line, 58
fuel burn-up, 35, 70
Gaia hypothesis, 116, 194
galactic center, 107, 111
galactic jets, 107, 108, 112
galaxy, 108, 109, 111
Galileo, 4, 111, 170
Gandhi, 3
garbage in, garbage out, 134
gas giants, 59, 95
gaseous protoplanet, 39, 60, 61, 62, 63, 64, 72, 75, 81, 95
Gauss, 11, 31, 171
genotoxicity, 125
geodynamics, 38, 41, 52, 54, 69, 178, 187
geology, 2, 41, 49, 52, 54, 69, 75, 169, 182
geomagnetic field, 11, 12, 13, 14, 25, 26, 30, 32, 33, 34, 35, 47, 49, 52, 54, 74, 77, 78, 117, 167, 168, 169, 175
geomagnetic field collapse, 35, 52, 77, 78, 168, 169
geomagnetic reversal, 13, 14, 34, 35, 79, 170, 181
geoneutrino, 176, 177
geophysics, 6, 28, 75, 80, 169, 176, 186, 210
georeactor, 25, 26, 27, 28, 29, 30, 31, 33, 34, 35, 37, 47, 48, 49, 52, 69, 70, 71, 72, 75, 76, 77, 78, 79, 93, 134, 167, 168, 176, 177, 178, 205
georeactor nuclear fission energy, 37, 79, 134
geothermal gradient, 47, 55, 79
Giant Sequoia, 198

Gilbert, 11, 31, 171, 173
glacial ice, 52, 132
global temperature, 133, 136
global warming, 131, 133, 134, 135, 137, 138, 141, 143, 145, 147, 148, 152, 156, 168, 169, 195, 199, 200, 201, 209
Gottschalk, 136, 137, 138, 202
governance, 3, 208
government funding, 2, 14
gravitational collapse, 99, 104
gravitational creep, 41, 54
greenhouse gases, 133, 134, 135, 136, 142, 168
Hahn, 100, 192
Harteck, 98, 191
Hayashi, 99, 192
heat sink, 33, 34, 71, 76
Hecker, 163
helium, 25, 26, 27, 47, 48, 49, 60, 69, 72, 73, 93, 98, 99, 107, 111, 112, 169, 175, 176
hexavalent chromium, 125
Hilgenberg, 37, 177
Hollenbach, 25, 28, 72, 169
homeostasis, 194
hot-spot, 72
How to Wreck the Environment, 128, 130, 199
Hubble Space Telescope, 65, 106, 107, 109
Hubble, Edwin P., 111
human nature, 2, 7, 8, 165
humanity, 10, 13, 34, 35, 36, 115, 117, 125, 163, 166, 195, 196, 200
Hunt, 139, 204
hydrogen, 1, 60, 63, 64, 91, 96, 99, 100, 101, 103, 107, 111, 112, 175, 191, 210
hydrogen bomb, 99, 100, 101,

103
ice crystal, 118, 119, 149
IceCube, 161, 162, 163, 165, 221
ices, 38, 39, 54, 58, 61, 75, 81, 94, 95, 96
immune response, 125
Indian Academy of Sciences, 31
infrastructure, 13, 35, 172
inner core, 13, 15, 16, 17, 18, 19, 20, 21, 22, 33, 34, 54, 72, 76, 167, 172
inner planets, 64, 65
insectivorous bats, 127
insects, 127, 156
Intergovernmental Panel on Climate Change, 134, 135, 136, 145, 202
iodine, 153
ionosphere heater, 35, 157, 159, 160, 165, 166
IPCC, 134, 135, 146, 202
Jet Propulsion Laboratory, 80
jet-emplaced, 119, 125, 126, 129
jet-laid, 117, 118, 128, 129
jet-sprayed, 118, 119, 120, 121, 122, 124, 145, 155
John 8
32, iii
Jupiter, 38, 39, 53, 54, 58, 65, 75, 81, 84, 102, 103, 172, 178, 191
Jupiter-like gas giant, 38, 53, 75, 178
Kamioka, 28, 31, 71
KamLAND, 28, 29, 176, 177
Kant, 60, 184
Kepler, 112
Kepler, Johannes, 112
Key West, 121
kimberlite, 40, 178
Kossmat, 44, 179
Kuroda, i, 101, 102, 192

Laplace, 60, 184
Large Igneous Province, 152, 187
Le Chatelier's Principle, 3
lead, 4, 52, 58, 60, 68, 73, 75, 77, 104, 111, 117, 125, 128, 147, 165, 174
Lehmann, i, 15, 16, 18, 19, 173
lies, iii, 14, 159
Lincoln, 28
Lockwood, 135
logical progression, 6, 7, 10, 20, 22, 25, 73
long-wave radiation, 143, 146
Lovelock, 116, 194, 201
lunar highlands, 67
lunar maria, 67, 68, 72, 187
lung cancer, 125, 147, 156, 197
MacDonald, 128, 130, 145, 157, 198, 199
magma ocean, 57, 183
magnesium, 17, 22, 23, 54, 72, 82, 83
magnesium sulfide, 22, 23
magnetic field, 11, 12, 31, 32, 69, 71, 72, 75, 76, 79, 85, 94, 95, 96, 103, 168, 171, 177, 189, 190, 205
Maher, 197
manganese, 154
mantle, 9, 14, 20, 22, 23, 24, 27, 33, 37, 39, 44, 46, 47, 48, 50, 54, 55, 61, 62, 69, 72, 74, 75, 79, 81, 95, 96, 139, 174, 175, 176, 179, 180, 187, 205
mantle convection, 9, 37, 44, 46, 55, 69, 174, 205
mantle decompression thermal tsunami, 47, 79, 95, 96
mantle plumes, 48, 176, 179
Mars, 58, 65, 68, 69, 81, 84, 85, 86, 87, 88, 89, 91, 92, 94, 95,

96, 189, 190, 191
mass extinction, 152, 181, 193, 211
mass ratio, 22, 23, 24, 183
Mercury, 62, 63, 64, 65, 69, 84, 91, 92, 96, 168, 183, 184
Mick West, 159, 160
micrograph, 21
microwave, 112
mid-ocean ridge system, 41, 42, 49
mobile aluminum, 120
Mohorovičić, 14
Molina, 149, 210
Montreal Protocol, 149, 150, 156
mortality, 125, 127, 147, 209
Moulton, 57, 182
Mount St. Helens, 142, 143, 206
mountain ranges, 40, 45, 75, 86, 170
Murrell, 24, 174
Nakano, 99, 192
NASA, 19, 62, 63, 67, 80, 86, 87, 88, 90, 102, 103, 128, 189, 193
NASA grant, 20
NASA MESSENGER, 63
National Academy of Sciences, 27
National Science Foundation, 2, 161
National Security Advisor, 162, 165
natural gas deposits, 50, 51, 55, 93, 180
Naturwissenschaften, 24, 102, 168, 174
neurodegenerative diseases, 125
New York Times, 136
nickel, 16, 17, 19, 20, 22, 39, 54, 68, 72, 167
nickel silicide, 17, 19, 20, 22, 54, 72, 167
nuclear chemistry, 1, 36
nuclear fission, 25, 28, 29, 30, 31, 33, 38, 41, 48, 54, 69, 70, 71, 72, 73, 75, 76, 79, 80, 81, 86, 94, 95, 96, 100, 101, 102, 103, 104, 107, 108, 112, 168, 169, 176
nuclear fission chain reaction, 48, 101, 103, 104, 107, 108, 112
nuclides, 41, 126
Oak Ridge National Laboratory, 25, 31, 48, 71, 72
ocean basins, 38, 39, 40, 41, 43, 52, 55
ocean floor, 37, 42, 94
Oklo, 102, 192
Oldham, 14, 171
Oliphant, 98, 191
ordinary chondrite, 16, 17, 20, 57, 58, 60, 68, 82, 83, 183
origin of the elements, 111
oxidative stress, 197
oxygen-starved, 17, 38, 54, 60, 62, 71, 76
ozone depletion, 148, 149, 150, 151, 152, 154, 155, 156, 196, 197, 209, 210, 215
ozone hole, 148, 149, 150, 151, 209, 210
ozone layer, 124, 148, 149, 156, 217
paleolatitude, 46, 75, 179
Pangaea, 9, 37, 46, 52
paradigm, 8, 9, 13, 36, 46, 52, 53, 68, 75, 81, 104, 135, 138, 139, 142, 145, 147, 148, 168, 169, 197
particulate pollution, 125, 133, 137, 138, 141, 143, 146, 147,

195, 198, 200
particulate trails, 126, 128, 129
Penzias, 112, 193
Permian, 152, 190, 211
perspective, 86, 185, 211
petroleum, 50, 51, 55, 180
phased-array, 160, 161, 162, 163, 165
phenomenological models, 36
photosphere, 58
Physical Review Letters, 28, 33
physically impossible, 8, 9, 32, 33, 44, 46, 69, 76, 139, 169
Physics Letters, 28, 176
pillow basalts, 40
planet migration, 59, 183
planetesimal theory, 57, 60, 68, 74, 84
plankton, 127, 197
plants, 115, 126, 156
plate tectonics, 9, 37, 44, 46, 50, 52, 55, 74, 75, 174
populations, 127, 156, 166, 196
premature death, 147
primordial, 27, 38, 39, 53, 54, 57, 58, 60, 62, 82, 94, 106, 112, 176
primordial condensation, 38, 58, 60, 62, 82
Proceedings of the National Academy of Sciences, 27, 181, 193, 211
Proceedings of the Royal Society of London, 35, 58, 104, 185
protoplanetary theory, 60, 68, 74, 84, 170
Ptolemaic, 8
pyrite, 52
pyroclastic, 152
Qu, 142, 205
radiation-balance, 135, 138

radioactive elements, 28, 95, 125
Raghavan, 28, 29, 176
rainwater, 119, 120, 121, 122, 155
reactor, 25, 26, 30, 38, 47, 54, 72, 73, 76, 93, 94, 96, 101, 102, 168, 169, 175, 176, 192, 215
red tide, 207
remanent magnetization, 72
Republic of Cyprus, 128
rheology, 41
Rio Grande Rift System, 51
rocky, 10, 24, 33, 38, 39, 54, 58, 95
Rowland, 149, 210
Rudee, 21, 173
Rutherford, 98, 191
Saharan, 144, 203, 206, 207, 216
salt water, 127
Scheidegger, 37
scientific fraud, 145
scientific truth, 13, 147
sea-level changes, 52, 181
seismologists, 23
shadow zone, 14, 15
Siberian Traps, 48, 51, 69, 72, 93, 94, 152, 187
silicon, 17, 54, 82, 83
Slipher, Vesto M., 111
smokestacks, 123, 152
snow, 121, 122, 123, 155
snow mold, 123, 155
solar system, 57, 58, 59, 74, 81, 84, 92, 116, 168, 170, 177, 182
species extinction, 40, 52, 115, 117, 166, 169
spherical particles, 124
spiral galaxy, 105, 108, 109
Strassmann, 100, 192
strontium, 119, 120

sub-core, 33, 71, 76
submarine canyons, 44, 45, 55, 69, 178
sub-shell, 33, 34, 35, 71, 76, 77, 78, 79
Suess, i, 1, 2, 4, 5, 17, 32, 44, 60, 61, 81, 167, 173, 183
Suess, Hans E., 4
Sullivan, 162, 165
supercontinent cycles, 9, 46, 75
supernova, 111
super-volcano, 35, 49, 180
Talukdar, 144, 206
Technology Bill of Rights, 166, 199
Teller, 99, 100, 104, 191
terawatts, 26, 31, 48, 70, 71
thermal convection, 32, 33, 71, 74, 76, 139, 140
thermal expansion, 37, 74, 139
thermodynamic considerations, 58, 60, 81
thermonuclear fusion, 38, 98, 99, 101, 107, 108, 111, 112
thermonuclear ignition, 95, 104, 108, 109, 113, 191
time machine, 6
tissue inflammation, 125
toxic, 116, 117, 120, 123, 124, 125, 126, 127, 159, 197, 217
trauma, 35, 77, 111
Trojan horse, 130
Tromsø, 159, 160, 219, 221
troposphere, 119, 120, 129, 137, 139, 141, 143, 144, 146, 152, 155, 217
troughs, 44, 55

truth, iii, 3, 36, 97, 113, 166
tsunami, 47, 95, 165
T-Tauri phase, 39, 54, 64, 95
Turkey earthquake, 165
ultraviolet radiation, 124, 126, 127, 152, 156
United Nations, 130, 133, 134, 135, 145, 149, 157, 199
Universe, 3, 8, 97, 106, 107, 111, 112, 113, 168, 170, 177, 220
University of Oxford, 30
Urey, i, 1, 2, 4, 5, 183, 184, 186
UV Index, 151
UV-B, 197
UV-C, 197
Valles Marineris, 86, 87, 88, 89, 90, 93, 96, 190
volcanic eruption, 40, 74, 117, 146, 169, 189, 206
war crimes, 163, 223
Wegener, 37, 177
West Siberian Basin, 51
white haze, 117
Whole-Earth Decompression Dynamics, 38, 39, 41, 43, 44, 45, 49, 50, 52, 53, 54, 55, 69, 75, 92, 95, 168, 170
Whole-Mars Decompression Dynamics, 69, 80, 86, 94, 95, 96, 170
wildfires, 117, 127, 156, 195
Wilson cycles, 9
World Health Organization, 145, 208
World War II, 36, 136, 137, 200, 202
Yellowstone, 49, 175, 179, 180

ABOUT THE AUTHOR

J. Marvin Herndon earned the BA degree in physics in 1970 from the University of California, San Diego, PhD degree in nuclear chemistry in 1974 from Texas A&M University, and received advanced training in geochemistry and cosmochemistry at the University of California, San Diego.

Dr. Herndon's published discoveries include: Recognizing that Earth's early formation as a Jupiter-like gas giant makes it possible to derive virtually all the geological and geodynamic behavior of our planet, origin of fold-mountain chains, primary initiation of fjords and submarine canyons; origin and typography of ocean floors and continents via his Whole Earth Decompression Dynamics; Earth's previously unanticipated and potentially variable energy and heat sources, such as a planetocentric nuclear fission reactor at the center of Earth; the reason why our Moon's two faces are strikingly different; origin of Earth's magnetic field and cause of magnetic reversals; new Martian paradigm called Whole Mars Decompression Dynamics; how stars, including our own sun, ignite; and the reason why the vast multitude of galaxies in the universe display just a few prominent patterns of luminous stars; and discovery that particulate pollution, not carbon dioxide, is the primary cause of anthropogenic global warming. His forensic scientific investigations in collaboration with Mark Whiteside, MD, MPH of jet-laid aerosol trails led to the discovery of the pseudo-legal basis of United Nations sanctioned environmental warfare, notably causing Arctic melting; and discovery that coal fly ash, not chlorofluorocarbons, is the principal cause of stratospheric ozone depletion..

Printed in Great Britain
by Amazon